REA: THE TEST PREP AP TEACHERS RECOMMEND

# AP* BIOLOGY
# ALL ACCESS™

**Amy Slack, Ph.D.**
AP Biology Teacher
The Westminster Schools
Atlanta, Georgia

**Melissa Kinard, Ph.D.**
Former AP Biology Teacher
Gwinnett County Public Schools
Atlanta, Georgia

 **Research & Education Association**
Visit our website: www.rea.com

**Planet Friendly Publishing**
✓ Made in the United States
✓ Printed on Recycled Paper
    Text: 10%    Cover: 10%
Learn more: www.greenedition.org

At REA we're committed to producing books in an Earth-friendly manner and to helping our customers make greener choices.

Manufacturing books in the United States ensures compliance with strict environmental laws and eliminates the need for international freight shipping, a major contributor to global air pollution.

And printing on recycled paper helps minimize our consumption of trees, water and fossil fuels. This book was printed on paper made with **10% post-consumer waste**. According to the Environmental Paper Network's Paper Calculator, by using these innovative papers instead of conventional papers, we achieved the following environmental benefits:

Trees Saved: 5 • Air Emissions Eliminated: 968 pounds
Water Saved: 1,095 gallons • Solid Waste Eliminated: 286 pounds

Courier Corporation, the manufacturer of this book, owns the Green Edition Trademark.
For more information on our environmental practices, please visit us online at www.rea.com/green

*Research & Education Association*
61 Ethel Road West
Piscataway, New Jersey 08854
E-mail: info@rea.com

## AP BIOLOGY ALL ACCESS™

**Copyright © 2014 by Research & Education Association, Inc.**
All rights reserved. No part of this book may be reproduced in any form without permission of the publisher.

Printed in the United States of America

Library of Congress Control Number 2013944516

ISBN-13: 978-0-7386-1081-8
ISBN-10: 0-7386-1081-X

LIMIT OF LIABILITY/DISCLAIMER OF WARRANTY: Publication of this work is for the purpose of test preparation and related use and subjects as set forth herein. While every effort has been made to achieve a work of high quality, neither Research & Education Association, Inc., nor the authors and other contributors of this work guarantee the accuracy or completeness of or assume any liability in connection with the information and opinions contained herein and in REA's software and/or online materials. REA and the authors and other contributors shall in no event be liable for any personal injury, property or other damages of any nature whatsoever, whether special, indirect, consequential or compensatory, directly or indirectly resulting from the publication, use or reliance upon this work.

All trademarks cited in this publication are the property of their respective owners.

Cover image: © KidStock/Getty Images

All Access™ and REA® are trademarks of Research & Education Association, Inc.

# Contents

About Research & Education Association .................................................................x
Acknowledgments .....................................................................................................x
About Our Authors ..................................................................................................xi

## Chapter 1: Welcome to REA's All Access for AP Biology — 1

## Chapter 2: Strategies for the Exam — 7

What Will I See on the AP Biology Exam? ............................................................... 7
What's the Score? ..................................................................................................... 9
Section I: Strategies for the Multiple-Choice and Grid-In Sections of the Exam ...... 10
Achieving Multiple-Choice Success ........................................................................ 12
The Grid-In Questions ........................................................................................... 15
Section II: Strategies for the Free-Response Section of the Exam ........................... 18
Achieving Free-Response Success ........................................................................... 20
A Sample Response ................................................................................................ 24

## Big Idea 1: The Process of Evolution Drives the Diversity and Unity of Life

## Chapter 3: Natural Selection and Evolution — 29

Contributing Ideas to Darwin's Descent with Modification .................................... 29
Natural Selection .................................................................................................... 30

## Chapter 4: Evolution: An Ongoing Process — 33

Population Genetics and Hardy-Weinberg Equilibrium ........................................... 33

Phenotypic Variation.................................................................................................. 36

Speciation ................................................................................................................... 40

Quiz 1 ........................................... *available online at www.rea.com/studycenter*

## Chapter 5: Common Ancestry and Models of Evolutionary History — 45

Common Ancestry..................................................................................................... 45

Models of Evolutionary History ................................................................................ 49

## Chapter 6: The Origin of Life — 53

Origin of Life—Hypothesis and Evidence................................................................ 53

Earth's History and the Origin of Life ...................................................................... 55

Extinction and Adaptive Radiation .......................................................................... 56

## Chapter 7: AP Biology Labs: Evolution — 59

Lab 1: Artificial Selection ........................................................................................... 59

Lab 2: Mathematical Modeling—Using Hardy-Weinberg ...................................... 62

Lab 3: Comparing DNA Sequences—BLAST Lab................................................... 64

Quiz 2 ........................................... *available online at www.rea.com/studycenter*

## Big Idea 2: Biological Systems Utilize Free Energy and Molecular Building Blocks to Grow, to Reproduce, and to Maintain Dynamic Homeostasis.

## Chapter 8: Energy and Matter — 67

Energy Changes–The Chemistry of Life ................................................................... 67

    Modes of Energy Capture .................................................................................... 72

    Matter................................................................................................................... 74

    Biological Chemistry............................................................................................ 74

    Water.................................................................................................................... 77

    Monomers and Polymers..................................................................................... 79

    Biological Molecules ........................................................................................... 80

    Enzymes............................................................................................................... 85

## Chapter 9: Photosynthesis and Cellular Respiration     87

    Photosynthesis ..................................................................................................... 87

    Fermentation and Cellular Respiration................................................................ 93

    Fermentation ....................................................................................................... 98

## Chapter 10: Cellular Structure, Membranes, and Transport     101

    Prokaryotes and Eukaryotes .............................................................................. 101

    The Endomembrane System and Eukaryotic Organelles................................... 103

    Cell Membrane Structure.................................................................................. 106

    Cell Membrane Transport ................................................................................ 109

    Quiz 3 .................................................... *available online at www.rea.com/studycenter*

## Chapter 11: Homeostasis     119

    Homeostasis in Plants and Animals................................................................... 119

    Coordinated Cooperation—Organelles............................................................. 119

    Cells, Tissues, and Organs................................................................................. 123

    Organ and Systems Interaction ......................................................................... 125

    Negative and Positive Feedback Mechanisms ................................................... 126

    Evolutionary Similarities and Examples............................................................ 130

    Disruptions of Homeostasis .............................................................................. 131

## Chapter 12: Reproduction, Growth, and Development — 135

Reproductive Process of All Land Plants ................................................................ 135

Plant Growth and Development .............................................................................. 136

Reproduction ............................................................................................................. 137

Animal Development ............................................................................................... 138

## Chapter 13: AP Biology Labs: Biological Processes and Energetics — 143

Lab 4: Diffusion and Osmosis ................................................................................. 143

Lab 5: Photosynthesis .............................................................................................. 149

Lab 6: Respiration .................................................................................................... 151

Quiz 4 ......................................................... *available online at www.rea.com/studycenter*

Mini-Test 1 ................................................. *available online at www.rea.com/studycenter*

## Big Idea 3: Living Systems Store, Retrieve, Transmit, and Respond to Information Essential to Life Processes

## Chapter 14: DNA Structure and Replication and RNA Structure and Gene Expression — 157

Key Concepts ............................................................................................................ 157

Discovery of DNA as the Genetic Material .......................................................... 157

DNA Structure ......................................................................................................... 159

DNA Replication ..................................................................................................... 163

RNA Compared to DNA ........................................................................................ 165

Transcription ............................................................................................................ 166

Translation ................................................................................................................ 167

## Chapter 15: Cell Cycle, Mitosis, and Meiosis — 173

Overview .................................................................................................. 173
The Cell Cycle........................................................................................... 173
Meiosis...................................................................................................... 177
The Chromosomal Basis of Inheritance .................................................... 181
**Quiz 5** ........................................... *available online at www.rea.com/studycenter*

## Chapter 16: Regulation of Gene Expression and Genetic Variation — 189

Overview .................................................................................................. 189
Gene Regulation in Prokaryotes ............................................................... 190
Gene Regulation in Eukaryotes ................................................................ 191
Applications of Genetic Engineering ........................................................ 194
Genotype Affects Phenotype .................................................................... 197
Viral Replication ....................................................................................... 199

## Chapter 17: Cell Communication — 203

Cell Communication ................................................................................ 203
Chemical Signaling ................................................................................... 205
Signal Transduction .................................................................................. 206

## Chapter 18: Information Transmission — 209

Information Exchange .............................................................................. 209
Cognition .................................................................................................. 211
Regulation of Behavior ............................................................................. 214

## Chapter 19: AP Biology Labs: Genetics — 219

Lab 7: Mitosis and Meiosis ........................................................................................ 219

Lab 8: Bacterial Transformation ............................................................................... 223

Lab 9: Restriction Enzyme Analysis ......................................................................... 225

Quiz 6 ................................................... *available online at www.rea.com/studycenter*

## Big Idea 4: Biological Systems Interact, and These Systems and Their Interactions Possess Complex Properties

## Chapter 20: Interactions — 229

Communities .............................................................................................................. 229

Movement of Matter and Energy ............................................................................. 231

## Chapter 21: Competition and Cooperation — 237

Molecular Interactions .............................................................................................. 237

Organism Interactions ............................................................................................... 242

Population Interactions ............................................................................................. 247

Distribution of Ecosystems ....................................................................................... 249

## Chapter 22: Environmental Interactions — 253

Molecular Variation .................................................................................................. 253

Environmental Effects on Genotype ....................................................................... 256

Population Dynamics ................................................................................................ 257

Biodiversity ................................................................................................................. 258

## Chapter 23: AP Biology Labs: Biological Interactions     263

    Lab 10: Energy Dynamics ................................................................... 263

    Lab 11: Transpiration ........................................................................ 267

    Lab 12: Fruit Fly Behavior ................................................................. 270

    Lab 13: Enzyme Action ..................................................................... 273

    **Quiz 7** ............................................. *available online at www.rea.com/studycenter*

    **Mini-Test 2** .................................... *available online at www.rea.com/studycenter*

## Practice Exam (also available online at *www.rea.com/studycenter*)     281

    Answer Key ...................................................................................... 315

    Detailed Explanations of Answers ..................................................... 316

## Answer Sheets     339

## Glossary     341

## Appendix: AP Biology Equations and Formulas     353

## Index     357

## About Research & Education Association

Founded in 1959, Research & Education Association (REA) is dedicated to publishing the finest and most effective educational materials—including study guides and test preps—for students in middle school, high school, college, graduate school, and beyond.

Today, REA's wide-ranging catalog is a leading resource for teachers, students, and professionals. Visit *www.rea.com* to see a complete listing of all our titles.

## Acknowledgments

We would like to thank Pam Weston, Publisher, for setting the quality standards for production integrity and managing the publication to completion; John Paul Cording, Vice President, Technology, for coordinating the design and development of the REA Study Center; Larry B. Kling, Vice President, Editorial, for his overall direction; Diane Goldschmidt, Managing Editor, for coordinating development of this edition; Jennifer Guercio for technical review and edit of the original manuscript; Marianne L'Abbate for copyediting; Transcend Creative Services for typesetting and indexing this edition; Ellen Gong for proofreading; and Weymouth Design and Christine Saul, Senior Graphic Designer, for designing our cover.

# About Our Authors

**Amy Slack, Ph.D.,** currently teaches AP Biology and Honors Biology at The Westminster Schools in Atlanta, Georgia. Ms. Slack earned her B.S. in Biology from Centre College and her M.S. in Biology from Georgia State University. Serving as a teaching assistant during graduate school prompted her to change her career path to education and to complete her M.Ed. and Ph.D. in Science Curriculum and Instruction. Ms. Slack has been published by the National Science Teacher's Association.

**Melissa Kinard, Ph.D.,** taught all levels of biology, including AP Biology, Physical Science and Chemistry in the Gwinnett County Public School system for more than 25 years. She earned her M.S. and Ph.D. in Science Education from Georgia State University. In 2007, Ms. Kinard was named Teacher of the Year by Duluth High School students.

# Chapter 1

# Welcome to REA's All Access for AP Biology

## A new, more effective way to prepare for your AP exam

There are many different ways to prepare for an AP exam. What's best for you depends on how much time you have to study and how comfortable you are with the subject matter. To score your highest, you need a system that can be customized to fit you: your schedule, your learning style, and your current level of knowledge.

This book, and the free online tools that come with it, will help you personalize your AP prep by testing your understanding, pinpointing your weaknesses, and delivering flashcard study materials unique to you.

Let's get started and see how this system works.

## How to Use REA's AP All Access

The REA AP All Access system allows you to create a personalized study plan through three simple steps: targeted review of exam content, assessment of your knowledge, and focused study in the topics where you need the most help.

Here's how it works:

| | |
|---|---|
| **Review the Book** | Study the topics tested on the AP exam and learn proven strategies that will help you tackle any question you may see on test day. |
| **Test Yourself & Get Feedback** | As you review the book, test yourself. Score reports from your free online tests and quizzes give you a fast way to pinpoint what you really know and what you should spend more time studying. |
| **Improve Your Score** | Armed with your score reports, you can personalize your study plan. Review the parts of the book where you are weakest, and use the REA Study Center to create your own unique e-flashcards, adding to the 100 free cards included with this book. |

## Finding Your Weaknesses: The REA Study Center

The best way to personalize your study plan and truly focus on your weaknesses is to get frequent feedback on what you know and what you don't. At the online REA Study Center, you can access three types of assessment: topic-level quizzes, mini-tests, and a full-length practice test. Each of these tools provides true-to-format questions and delivers a detailed score report that follows the topics set by the College Board.

### Topic-Level Quizzes

Short online quizzes are available throughout the review and are designed to test your immediate grasp of the topics just covered.

### Mini-Tests

Two online mini-tests cover what you've studied in each half of the book. These tests are like the actual AP exam, only shorter, and will help you evaluate your overall understanding of the subject.

### Full-Length Practice Test

After you've finished reviewing the book, take our full-length exam to practice under test-day conditions. Available both in this book and online, this test gives you the most complete picture of your strengths and weaknesses. We strongly recommend that you take the online version of the exam for the added benefits of timed testing, automatic scoring and a detailed score report.

## Improving Your Score: e-Flashcards

Once you get your score reports from the online quizzes and tests, you'll be able to see exactly which topics you need to review. Use this information to create your own flashcards for the areas where you are weak. And, because you will create these flashcards through the REA Study Center, you'll be able to access them from any computer or smartphone.

Not quite sure what to put on your flashcards? Start with the 100 free cards included when you buy this book.

## After the Full-Length Practice Test: *Crash Course*

After finishing this book and taking our full-length practice exam, pick up REA's *Crash Course for AP Biology*. Use your most recent score reports to identify any areas where you are still weak, and turn to the *Crash Course* for a rapid review presented in a concise outline style.

# REA's Suggested 8-Week AP Study Plan

Depending on how much time you have until test day, you can expand or condense our eight-week study plan as you see fit.

To score your highest, use our suggested study plan and customize it to fit your schedule, targeting the areas where you need the most review.

|  | Review<br>1-2 hours | Quiz<br>15 minutes | e-Flashcards<br>Anytime, Anywhere | Mini-Test<br>30 minutes | Full-Length Practice Test<br>3 hours |
|---|---|---|---|---|---|
| Week 1 | Chapters 1–4 | Quiz 1 | Access your e-flashcards from your computer or smartphone whenever you have a few extra minutes to study.<br><br>Start with the 100 free cards included when you buy this book. Personalize your prep by creating your own cards for topics where you need extra study. | | |
| Week 2 | Chapters 5–7 | Quiz 2 | | | |
| Week 3 | Chapters 8–10 | Quiz 3 | | | |
| Week 4 | Chapters 11–13 | Quiz 4 | | Mini-Test 1 (The Mid-Term) | |
| Week 5 | Chapters 14–15 | Quiz 5 | | | |
| Week 6 | Chapters 16–19 | Quiz 6 | | | |
| Week 7 | Chapters 20–23 | Quiz 7 | | Mini-Test 2 (The Final) | |
| Week 8 | Review Chapter 2 Strategies | | | | Full-Length Practice Exam (Just like test day) |

*Need even more review?* Pick up a copy of REA's *Crash Course for AP Biology,* a rapid review presented in a concise outline style. Get more information about the *Crash Course* series at *www.rea.com.*

## Test-Day Checklist

| | |
|---|---|
| ✓ | Get a good night's sleep. You perform better when you're not tired. |
| ✓ | Wake up early. |
| ✓ | Dress comfortably. You'll be testing for hours, so wear something casual and layered. |
| ✓ | Eat a good breakfast. |
| ✓ | Bring these items to the test center:<br>• Several sharpened No. 2 pencils<br>• Admission ticket<br>• Two pieces of ID (one with a recent photo and your signature)<br>• A noiseless wristwatch to help pace yourself. |
| ✓ | Arrive at the test center early. You will not be allowed in after the test has begun. |
| ✓ | Relax and compose your thoughts before the test begins. |

**Remember: eating, drinking, smoking, cellphones, dictionaries, textbooks, notebooks, briefcases, and packages are all prohibited in the test center.**

# Strategies for the Exam

Chapter 2

The AP Biology curriculum is divided into four Big Ideas:

1. **Evolution**—The evolutionary process is responsible for the diversity of life.

2. **Cellular processes: energy and communication**—Biological systems use molecular building blocks and energy to maintain homeostasis, reproduce, and grow.

3. **Genetics and information transfer**—Living systems store, retrieve, transmit and respond to information essential to life processes.

4. **Interactions**—Biological systems interact and possess complex properties.

All of these Big Ideas are covered in this book, but what will the actual exam look like?

## What Will I See on the AP Biology Exam?

The AP Biology exam is made up of two sections: multiple-choice and free-response. Each section includes questions that test students' understanding of the 4 Big Ideas.

### The Multiple-Choice Section

The multiple-choice section requires you to answer 63 multiple-choice questions and 6 grid-in questions in 90 minutes. This section tests your ability to not just remember facts about the various fields of biology, but also to synthesize that knowledge in new scenarios and to interpret and analyze data from experiments.

Here are the major fields of inquiry covered on the AP Biology exam:

- Evolution
- Cellular structure
- Energetics
- Genetics
- Communication
- Science practices

Being able to name which organelle converts light energy into chemical energy (the chloroplast, but you know that, right?) will not do you much good unless you can also explain how living organisms use this chemical energy to carry out life's functions and discuss how energy is transformed as it moves through cells to organisms and to ecosystems. It sounds like a lot to handle from multiple-choice questions, but by working efficiently and methodically you'll have plenty of time to answer the questions in this section. We'll look at this in greater depth later in this chapter.

## The Free-Response Section

After time is called on the multiple-choice section, you'll get a short break before diving into the free-response section. This section requires you to produce eight written responses in 80 minutes. You will have a mandatory 10-minute reading period to review the questions, but you can't start answering them. Like the multiple-choice section, the free-response portion of the exam expects you be able to apply your own knowledge to analyze biological information and data, in addition to being able to provide essential facts and definitions.

## A Word About the AP Biology Labs

All AP Biology students are expected to learn six Science Practices, which include reading and interpreting graphs, data analysis, and mathematical calculations. The Science Practices are covered in both the content and in the labs. In the AP Biology curriculum, there are 13 AP Biology labs divided up by the Big Ideas. The College Board recommends that students complete 8 labs, two from each of the four Big Ideas. You may have done more than 8 in your AP Biology class, but it is unlikely that you conducted all 13 labs. Because no student will have completed all the labs, the AP exam is not likely to ask specific details from each lab. Instead, the labs were designed for you to learn to ask questions, collect and analyze data, and draw conclusions.

The labs are organized around the theme of "inquiry." This means that you probably learned how to perform a technique, like building a microrespirometer, then used that technique to answer a question of your own design, such as "do crickets or mealworms have a higher respiration rate?" Because each AP Biology student conducts slightly different experiments, the lab sections in this book present possible data collected instead of every possible scenario. It is best to read each lab section, understand the concepts behind each lab, and make sure you can interpret and analyze the data presented. Additionally, if mathematical calculations are required, those are also discussed. Be certain you know how to analyze the data mathematically—for example, by calculating the rate of a reaction.

In this book, the AP labs are discussed in a chapter found at the end of each section. The chapters explain the lab concepts, science practices and data analysis related to each of the 13 Biology labs.

## What's the Score?

Although the scoring process for the AP exam may seem quite complex, it boils down to two simple components: your multiple-choice score plus your free-response scores. The multiple-choice section accounts for one-half of your overall score. There is no deduction for incorrect answers or questions left blank, so go ahead and take an educated guess! The free-response section also accounts for one-half of your total score. Trained graders, called readers, will read students' written responses and assign points according to grading rubrics. The number of points you accrue out of the total possible will form your score on the free-response section.

The test developer awards AP scores on a scale of 1 to 5. Although individual colleges and universities determine what credit or advanced placement, if any, is awarded to students at each score level, these are the assessments typically associated with each numeric score:

5   Extremely well qualified

4   Well qualified

3   Qualified

2   Possibly qualified

1   No recommendation

# Section I: Strategies for the Multiple-Choice and Grid-In Sections of the Exam

## Multiple-Choice Questions

Because the AP exam is a standardized test, each version of the test from year to year must share similarities in order to be fair. That means that you can always expect certain things to be true about your AP Biology exam.

Which of the following phrases is NOT an accurate description of a multiple-choice question on the AP Biology exam?

(A) always has four choices

(B) may have "all of the above" or "none of the above" as answer choices

(C) may ask you to find a wrong idea or to group related concepts

(D) may rely on a data table, diagram, or other visual

> Did you pick "B"? If so, good job! There will NOT be answer choices stating "all of the above" or "none of the above" on the exam.

So how you do study for the exam? You should focus on the application and interpretation of the various topics of biology rather than on the nuts and bolts. There's no need for you to memorize the steps and enzymes of the citric acid cycle. Keep in mind, too, that many biological concepts overlap—think of the Big Ideas! This means that you should consider the connections among ideas and concepts as you study. This will help you prepare for more difficult interpretation questions and give you a head start on questions that ask you to use Roman numerals to organize ideas into categories. Not sure what a typical Roman-numeral type question looks like? Let's examine one:

Which of the following is used by cells to regulate gene expression?

  I. Alternative RNA splicing

 II. Maintenance of homeostasis.

III. Increase in positive feedback

 IV. Interference by miRNAs

(A) I only

(B) III and IV

(C) II only

(D) I and IV

> Take a moment to look over the answer choices before evaluating each Roman-numeral statement. Notice that numerals I and IV appear in more answer choices than the other roman-numeral statements. Evaluating those first can save you time. For example, if I is false, the only possible correct answers are (B) and (C). Conversely, if I is true, then you can eliminate choices (B) and (C). Remember that the correct answer will include ALL of the applicable Roman numerals. Did you pick (D)? That's the right answer!

## Types of Questions

You've already seen the Big Ideas and a list of the general content areas that you'll encounter on the AP Biology exam. But how do those different areas translate into questions or grid-ins?

| Question Type | Sample Question Stems |
| --- | --- |
| Charts, Graphs, and Tables | Examine the table of differences in nucleotide sequences and determine which organisms are most closely related. |
| Factual | What is the mostly likely source of genetic variation in …? |
| Interpretation of Data | Based upon the data presented in the graph, determine the reaction that proceeded at the greatest rate. |
| Diagrams | Which of the following diagrams most accurately demonstrates the synthesis of a new molecule of DNA through replication? |
| Calculations (grid-ins) | A population has 25% of its individuals expressing the recessive phenotype. Calculate the frequency of the dominant allele. |

# Achieving Multiple-Choice Success

It is true that you don't have a lot of time to complete the multiple-choice section of the AP Biology exam. It's also true that you don't need to get *every* question right in order to achieve a great score. Answering just two-thirds of the questions correctly—along with a good showing on the free-response section—can earn you a score of a 4 or 5.

By working quickly and methodically, however, you'll have all the time you'll need. If a multiple-choice question has a short stem with little information given, plan to spend about 40 seconds on it. Some questions may involve reading 4 to 5 sentences as well as examining a data table, a diagram, or a graph. They may have 2 or 3 multiple-choice questions associated with them. You will, of course, need to spend more time on these questions. It is also important that you read the question *carefully* to determine exactly what is being asked.

Let's look at some other strategies for answering multiple-choice items.

## Process of Elimination

You have probably used this strategy, intentionally or unintentionally, throughout your entire test-taking career. The process of elimination requires that you read each answer choice and consider whether it is the best response to the question given. Because the AP exam typically asks you to find the *best* answer rather than the only answer, it is always advantageous to read each answer choice. More than one choice may have some grain of truth to it, but one—the right answer—will be the most correct. Let's examine a multiple-choice question and use the process-of-elimination approach:

Wobble of the anticodon refers to the

(A) movement of mRNA out of the nucleus

(B) inaccuracy of base pairing, i.e., adenine is paired with guanine

(C) freedom in the pairing of the third base of the codon

(D) fact that the codons UUG and CUG both code for leucine

> To use the process of elimination, consider each option. Choices (A) and (B) have no relation to the anticodon, so they can be eliminated immediately. Cross out each choice as you eliminate it. Then consider the definitions of the remaining choices. Which most directly relates to wobble? Both (C) and (D) are correct statements, but only one refers to wobble. If you're unsure, you can return to the question later or just guess. You've got a 50% chance of being right! The correct answer is (C), because wobble refers to the fact that the third base in a codon can often be changed and still result in the same amino acid. While (D) is true, it is a change in the first base, not the third, so it's not the best (nor the correct) answer.

## Predicting

Testing each and every answer choice is another way of answering test questions, but it can be a time-consuming process. You may find it helpful to instead try predicting the right answer before you read the answer choices. For example, you know that the answer to the math problem two-plus-two will always be four. If you saw this multiple-choice item on a math test, you wouldn't need to systematically test each response, but could go directly to the correct answer. You can apply a similar technique to complex items on the AP exam by brainstorming your own answer to the question before reading the answer choices. Then, simply pick the answer choice closest to the one you brainstormed. Let's look at how this technique could work on a common type of question on the AP Biology exam—one with a visual stimulus.

A graph question will most likely never simply ask you to pinpoint a particular data point. Instead, you will need to understand trends in the graph and maybe even calculate the rate of a reaction. Use your knowledge of data analysis to make predictions in order to answer these types of questions. Always take a moment to study the graph before diving into the question.

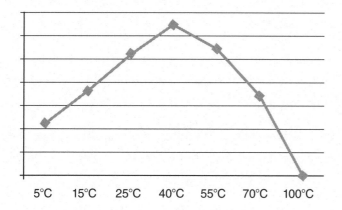

Look at the graph – this one doesn't have all the variables or a title. But, without reading the question or being given variables, you can tell a few things. Notice that it shows an increase in the dependent variable (on the *y*-axis) until it peaks at the middle condition of the independent variable (the *x*-axis). After peaking, the dependent variable then decreases all the way to 0.

Now read the question and see if you can predict the response.

The enzyme whose function is graphed above is most likely found in what type of organism?

(A) a thermophilic bacteria living in a deep sea vent

(B) a mammal living in the grasslands of the American Midwest

(C) a fish living in the cold waters of the Arctic Ocean

(D) an insect living in the Saharan desert

Which organism did you predict the answer would be?

You should have noticed from the graph that this enzyme had its optimal function at around 40°C. Hopefully you remembered this is approximately the temperature that humans maintain and thus went right to choice (B), because humans are mammals.

What should you do if you don't see your prediction among the answer choices? Apply the process of elimination to the remaining options to further home in on the right answer. Then, use your biology knowledge to make an educated guess.

Learning to predict takes some practice. You're probably used to immediately reading all of the answer choices for a question, but in order to predict well, you need to avoid doing this. Remember, the test developer doesn't want to make the right answers too obvious, so the wrong answers are intended to sound like appealing choices. You may find it helpful to cover up the answer choices to a question as you practice predicting. This will ensure that you don't sneak a peek at the choices too soon.

## Avoiding Common Errors

Answering questions correctly is always the goal on any exam, and one way to do that is to avoid these common mistakes:

- Missing key words that change the meaning of a question, such as *not, except*, or *least*. Circling these words in your test booklet will help you tune into them when answering the question.

- Overthinking an item and spending too much time agonizing over the correct response. If you can't answer the question or make an educated guess, move on. You may have time to come back to it.

- Changing your answer, but not completely erasing your first choice. The multiple-choice question section is scored by machine, so be certain to erase your original choice as thoroughly as possible.

- Falling for distractors. If you see a word that doesn't look familiar, it may be a distractor designed to trick you into choosing that answer.

## Some More Advice

Let's review what you've learned about answering multiple-choice questions. Using these techniques on practice tests will help you become comfortable with them before diving into the real exam, so be sure to apply these ideas as you work through this book.

- Big ideas are more important than minutiae. Focus on learning important concepts, models, and theories instead of memorizing facts and vocabulary.

- You only have a short period of time to complete each multiple-choice question, which may involve a lot of background reading. Pacing yourself during the quizzes, mini-tests, and practice exam that accompany this book can help you get used to these time constraints.

- Remember, there is no penalty for guessing, and making an educated guess is to your benefit. Use the process of elimination to narrow your choices. You might just guess the correct answer!

- Instead of spending valuable time pondering narrow distinctions or questioning your first answer, trust yourself to make good guesses most of the time.

- Read the question and predict what your answer would be before reading the answer choices.

- Expect the unexpected. You will see questions that ask you to apply information in various ways, such as reading about a new experiment or interpreting data you've never studied.

## The Grid-In Questions

The second part of the multiple-choice section of the AP Biology exam contains 6 grid-in questions. These are math questions that require you to perform simple calculations. Most of the questions will have graphs, charts, or tables of data to use in order to perform the calculations. There will not be a lot of background reading necessary to answer the questions. Read the question carefully and be certain you know exactly what the question is asking you to calculate. Pay attention to details: Is the question asking for a percent or a frequency? Is your scientific notation correct? You can do all the calculations correctly and still get the question wrong because of decimal placement.

You are allowed to use a four-function calculator on this exam. Most of the math could be solved without using the calculator, but it might slow you down. So be sure to bring a calculator to the exam (graphing calculators are not allowed).

The grid-in answer box will look like the diagram below. You can start from either the left or the right side to fill in your answer. Note that the grid has options for negative answers, decimals, and fractions. Be sure you choose the correct format of the answer. The question is likely to tell you to calculate to either a whole number, a tenth, or a hundredth.

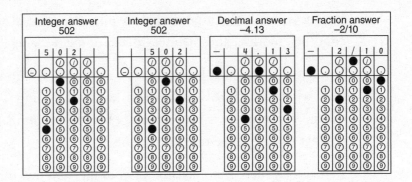

Topics for grid-in questions could include the following:

- Calculation of rate
- Chi square statistical analysis
- Hardy-Weinberg calculation
- Transfer of energy through an ecosystem
- Water potential

When you take your test, you will be given a formula sheet. The most important formulas will likely be the water potential, Chi square, and Hardy-Weinberg equations. Other items found on the formula sheet include:

- Probability
- Standard error, standard deviation, mean, mode, median, and range
- Solute potential
- Population growth rates
- Primary productivity
- Temperature coefficient
- Surface area and volume

- pH
- Gibbs Free Energy
- Dilution of a solution
- Metric prefixes

You won't need all of these equations for the test, but remember they are provided for you to use. If you forget what the symbol µ means, don't panic. Refer to the formula sheet and it will show you that µ means micro and is $10^{-6}$. Take some time to familiarize yourself with the formula sheet. If you know the location of the formulas and the format of the sheet, you'll be able to use it much more efficiently, and save yourself time on exam day.

Now, let's look at one example of a potential grid-in question:

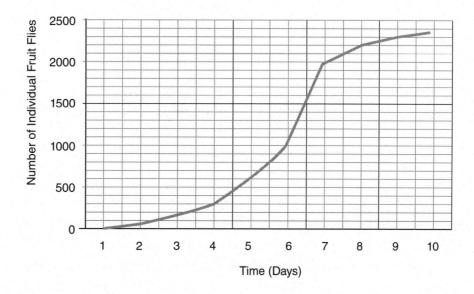

The graph above shows the growth in a population of fruit flies over 10 days. Calculate the rate of growth between days 3 and 7. Give your answer to the nearest whole number.

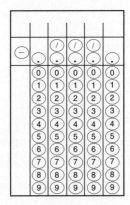

Did you get 462? Does your grid-in answer box match the example below? If so, great job!

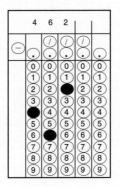

Remember that *rate* for any graph is the change in the *y*-axis over time. If you forget that, you could always refer to your formula sheet, which shows that population growth rate = $dY/dt$ where $dY$ = amount of change and $t$ = time. So, for the sample question, on day 7 there were 2000 fruit flies and on day 3 there were 150 fruit flies. This was a four-day period so: 2000 fruit flies—150 fruit flies/4 days = 462.5 fruit flies per day. Since there's no such thing as half a fruit fly, you must round either up or down. For this question, both 462 and 463 would be acceptable answers. See, that wasn't hard!

# Section II: Strategies for the Free-Response Section of the Exam

In the free-response section of the AP Biology exam, you will have 80 minutes to respond to two long and six short questions. The section begins with a mandatory 10-minute reading period to look over all the questions and think about your strategies to answer them. Often, these questions provide you with one or more visual stimuli, such as tables, diagrams, and graphs that might present experimental data.

The long free-response questions typically ask a series of increasingly sophisticated questions (usually in a sequence of four). The question might begin by asking you to read about a scientific experiment and interpret a data table. Then, you may be asked to graph the data and explain the results of the experiment. The final step might be predicting the outcome of the experiment if a variable is changed. This means that the long free-response questions typically build in difficulty within themselves.

The short free-response questions usually contain one to two embedded questions. These may ask you to compare and contrast two concepts—such as the humoral response versus the cell-mediated response in the immune system; to explain or draw a graph; or to describe a cellular process or mechanism, such as the importance of cell signaling in the nervous system.

Students with a deeper understanding of the content tested in the item will normally receive higher scores on these items than students with a superficial knowledge of the content. Expect at least one of the free-response questions to require you to combine knowledge of different content areas—for example, genetics and evolution or ecology and natural selection—in order to fully respond.

Although it's tempting to think of the free-response section as the essay section, that is not exactly correct. Unlike many other AP exams, you will not need to write a formal essay with an introduction and conclusion to answer the free-response questions on the AP Biology exam. It also does not mean that you will simply write a bulleted list of facts as your written answer to a free-response question. Instead, you'll need to write complete, coherent sentences that provide specific information requested by the various parts of a free-response question. Let's examine a typical free-response question.

## Sample Free-Response Question

You place yeast into an apparatus that can be used to count the bubbles of gas produced by the yeast cells. You give the yeast a solution containing glucose, which is needed for cellular respiration. The data you collected is in the table below.

> A text stimulus will precede the actual questions you must answer. This may provide additional information helpful in answering the questions.

| Temperature (°C) | 0 | 10 | 20 | 30 | 40 | 50 | 60 | 70 |
|---|---|---|---|---|---|---|---|---|
| Number of bubbles of gas produced per minute | 0 | 4 | 8 | 14 | 8 | 3 | 2 | 0 |

(a) **Graph** the results below. **Determine** the temperature that is optimum for yeast respiration.

(b) Respiration consists of a series of chemical reactions that are catalyzed by many enzymes. Considering what you know about respiration and enzymes, **analyze** and **explain** the experimental results.

> Many free-response questions provide a graph, table, chart, or other visual stimulus. Study and interpret this as you construct your response.

(c) **Design** an experiment that examines the effect of varying the pH of the glucose solution on the rate of respiration.

(d) **Predict** the expected results from part (c) by creating a graph.

Many free-response questions begin by asking you to plot data or read a graph. If graphing, be sure to label all your axes and clearly mark your data points.

As previously discussed, question parts will build in difficulty throughout the free-response item. Note that this question begins by asking you to draw a graph and read it, and it ends by asking you to predict the results of a second experiment of your own design. This not only requires that you draw a graph, but it also requires that you synthesize how a new experimental condition will affect the results. Be sure to number or letter the parts of your written response to help the AP readers follow your thinking.

## Achieving Free-Response Success

The single most important thing you can do to score well on the free-response section of the exam is to answer the questions that you are asked. It seems silly to point that out, doesn't it? But if you have ever written an essay and received a mediocre grade because you didn't fully answer the question asked, you know how easy it can be to stray off topic or neglect to include all the facts needed in a written response. By answering each of the free-response questions completely, you'll be well on your way to a great score on the AP exam. Let's look at some strategies to help you do just that.

- **Organize Your Time**

    You will have 80 minutes to answer all eight free-response questions. You may choose to spend as long as you like on each individual response. During the 10-minute mandatory reading period, read each question and consider whether any of them seem especially difficult or easy to you. You can then plan to spend more time addressing the more difficult items.

    Also remember that you may answer the questions in any order you wish. You may find it tempting to answer the simpler questions first to get them out of the way, but answering the more difficult questions first is usually a better use of your time. Why? Answering a difficult item first will make you feel less pressed for time, and getting it out of the way will be a relief. You'll be freshest on your first answer, so dealing with easier items when you're getting tired toward the end of the free-response section will be more manageable.

    In a similar strategy, you might want to answer the long free-response questions first. Budgeting your time can come in handy here, too. Knowing the material to a given item especially well will probably mean it takes less time for you to write your response. Don't be concerned that you are not spending enough time on a given question if you know that you have written a good, thorough answer. You are being scored on content, not effort!

- **Prewriting and Outlining**

    Yes, you have to write answers for 8 questions in 80 minutes on the AP Biology exam.

    Yes, that allows about 20 minutes to answer each of the long free-response questions and about 6 minutes for each of the short free-response questions.

    Yes, that seems like very little time—and it is…

    So your best strategy is to jump in and start writing, right? Wrong.

    If you spend four to five minutes on prewriting and outlining your answers to the long questions and one to two minutes outlining your answers to the short questions, your written responses will likely be clearer, complete, and quicker and easier for you to actually write. That's because creating a simple outline or bulleted list allows you to organize your thoughts, brainstorm good examples, and reject ideas that don't really work once you think about them. Use the structure of the free-response question to help build a quick outline.

Let's look at the previous free-response question and make a bulleted list of the tasks you need to do to answer the question.

- Make a graph
- Read the graph to determine an optimum condition (implies you're looking for a peak)
- Analyze the graph
- Explain why the graph shows this pattern
- Design an experiment (include a hypothesis, variables, and control)
- Predict the outcome—requires a graph and an estimation of potential data

Divide your outline into the same parts as the question stem, then write concise statements that directly answer each section of the question. You have essentially made a checklist of items to complete in order to finish this question. As you complete one part, cross it off your bulleted list. This way you're guaranteeing that you've answered all parts of this long free-response question.

- **Stick to the Topic**

Once you have written a good list or outline, stick to it! As you write your response, you'll find that most of the hard work is already done, and you can focus on expressing your ideas clearly, concisely, and completely. Don't include extra information that doesn't help you answer the question asked and only the question asked. AP readers will not award you extra points for adding lots of irrelevant information or giving personal anecdotes about your experiences in biology class.

Remember, too, that the readers know what information has been provided in the stimulus. If a free-response question contains a chart, graph, or diagram, don't waste your time and effort describing the contents of the visual stimulus. For example, in our example question, there is no need to restate that yeast produced 4 bubbles at 10 °C and 8 bubbles at 20 °C. It's given in the data table. Instead, recognize that the optimal condition for yeast respiration occurs at 20 °C because that's when the most carbon dioxide is produced, indicating that respiration is occurring at the greatest rate at that temperature.

- **Make It Easy on the AP Readers**

    As you are writing your responses, keep in mind what the AP readers will see when they sit down to consider your answers weeks from now. Expressing your ideas clearly and succinctly will help them know that you understand biology content and can apply it to new situations, thus ensuring that you will receive the best possible score. Using your clearest handwriting will also do wonders for your overall score; free-response graders are used to reading poor handwriting, but that doesn't mean they can decipher every scribble you might make. Printing your answers instead of writing them in cursive will make them easier to read, as will skipping lines in your answer booklet.

    Another good way to help the readers is to clearly label your answers according to the specific parts of the questions. Most questions are divided up into (a) and (b) parts. When you move on to the next part, label it clearly. If you draw a diagram, label it and explain it. If you are asked to draw a graph, label the axes and give it a descriptive title. Adding labels to each part of your response will help the AP readers follow your response through the multiple parts of a free-response question, and can only help your score. When you move on to the next question, clearly write the question number at the top of the page. It seems obvious, but you'd be surprised at the number of papers where the AP readers have to search for answers.

- **Revise in Three Minutes or Less**

    Even the best writers make mistakes, especially when writing quickly. Skipping or repeating words, misspelling complicated vocabulary words, or neglecting to include an important point from an outline are all common errors that occur when a test-taker is rushed. Reserving a few minutes at the end of your writing period will let you review your responses and make any necessary corrections. Adding skipped words or including forgotten information are the two most important edits you can make to your writing, because these will clarify your ideas and help your score. If it's a particularly complex vocabulary word, as so many in biology are (photophosphorylation, anyone?), don't sweat it, just do your best and make sure you explain what it means.

    The free-response readers are not mind readers. They will grade only what is on the page, not what you thought you were writing. Remember, too, that AP readers do not deduct points for wrong information, so you do not need to spend time erasing errors. Writing a sentence at the end of your answer or, if you've skipped lines, on the line below, corrects your mistake. Also, if you use a vocabulary word, think about including a definition with it. You may use the wrong word in your explanation, but have the correct concept, which might earn you a point.

## A Sample Response

After you have read, considered, outlined, planned, written, and revised, what do you have? A thoughtful free-response answer that is likely to earn you a good score, that's what. Of course, free-response readers must grade consistently in order for the test to be fair. All AP readers look for the same ideas in each answer to the same question by using a rubric. A rubric does not provide word for word explanations, but are instead bullet points of items that will earn a student points. Readers know their content and know if you're correct or not.

The long free-response questions are worth ten points each and the short free-response questions are worth two to four points each. The more points you earn, the better your overall score.

Let's examine how an AP reader might apply these grading techniques to a response to our sample question.

(a) **Graph** the results below. **Determine** the temperature that is optimum for yeast respiration.

*Sample student response:*

Optimum yeast respiration occurs at 30 °C because that is the peak of the graph when the most bubbles are formed.

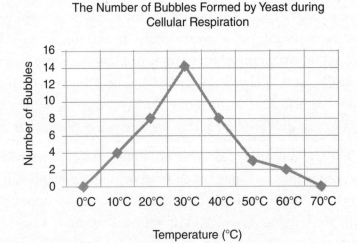

**Analysis:**

An AP reader looks for a clear and labeled graph. This student's graph included a title and labeled axes and would earn a point. A reader also needs to see a clear line with the correct data points, which earns a second point. Additionally, the student correctly identified the optimum temperature as 30 °C, earning a third point.

> (b) Respiration consists of a series of chemical reactions that are catalyzed by many enzymes. Considering what you know about respiration and enzymes, **analyze** and **explain** the experimental results.

*Sample student response:*

Cellular respiration increased as the temperature approached 30 °C, then decreased after reaching that maximum. The reason the reaction decreased as the temperature increased past optimum is because enzymes denature at high temperatures. Denaturing is the unfolding of an enzyme from its specific three-dimensional shape. If an enzyme doesn't have the right shape, then its active site shape will no longer be a match for the substrate, and the enzyme won't be able to bind to the substrate and catalyze the reaction. The reaction increased as it approached 30 °C because the increase in temperature provides more kinetic energy for the molecules, making it more likely that the enzyme and substrate will come into contact with one another.

**Analysis:**

Providing an analysis of the data (it increased, then decreased) is required to earn a point. This student's analysis was complete and earned one point. More points are earned for the explanation of the results. Explaining enzyme denaturing and kinetic energy earned this student two more points. A final point was earned for elaborating the explanation and explaining that the active site must match the substrate.

> (c) **Design** an experiment that examines the effect of varying the pH of the glucose solution on the rate of respiration.

*Sample student response:*

To determine the effect of various pH on the rate of respiration in yeast cells, the same apparatus as in the previous experiment should be used to count bubbles. The hypothesis is: if the pH of the glucose solution is varied between 2 and 10, then the yeast will respire at different rates, with 7 being the optimum pH. To begin this experiment, five solutions of glucose will be prepared having the following pHs: 2, 5, 7, 8, 10. The control is the pH of 7 because that is neutral. The pH is the independent variable and the dependent variable will be the number of carbon dioxide bubbles formed. Because carbon dioxide is a product of cellular respiration, it provides a way to quantify how fast respiration is happening in these yeast. Also all conditions of the experiment must be the same with each tube of yeast so the results are only because of the change in pH. This means that all the yeast must stay at the same temperature, around 30 °C, and receive the same amount of glucose.

**Analysis:**

Note that this answer isn't going to win any prizes for English literature. However, it's short, to the point, and discusses every part of a good experimental design. This student would earn four points for this answer, one point for each of the following elements: (1) a control, (2) an independent variable, (3) a measurable dependent variable, and (4) maintaining constant conditions except for the experimental variable.

(d) **Predict** the expected results from part (c) by creating a graph.

*Sample student response:*

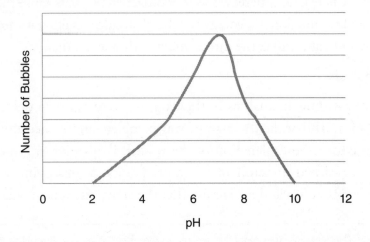

**Analysis:**

Finally, the student must demonstrate that he or she truly understands the effect of pH on enzyme function. The reader isn't looking for exact data points in this case or even an explanation; instead, the reader needs to see that the student predicts a near-neutral pH as optimum. The student earns an additional point for this simple sketch of a graph.

## Let's Review

Keep these pointers in mind when writing your answers to free-response questions:

- The shorter free-response questions will be similar to the longer free-response questions, but they are more likely to be specific to one topic. Use the same techniques to answer them as explained for the long free-response questions.

- You don't have to spend the suggested amount of time on each question. One question may be very straightforward and you may not need the entire time to answer it. If you know the answer, write your response and move on.

- Use your 10-minute reading period wisely and look at all the free-response questions before you start to write. You might want to answer the hardest question first and leave any easier questions for later.

- Make a clear and concise outline before you begin writing. This will help you organize your thoughts and speed up the actual writing process.

- Stay on topic and answer the question! Addressing the question fully is the single most important way to earn points on this section.

- Handwriting is important! If the AP reader can't read your writing, you'll get no points, even if your response is correct.

- Be sure to label the parts of your responses. Make it easy for the AP reader to award you points by being able to easily navigate your response. If you make it clear and easy for the reader, you'll earn the reward!

- Set aside a few minutes to review and revise your answers. You don't need to check the spelling of every word, but you do need to make sure that all of your ideas make it onto the page. Skipping lines while you write will leave room for you to add important words and ideas, and make it easier for the scorer to read your handwriting.

## Two Final Words: Don't Panic!

The free-response questions can and probably will ask you about specific biological concepts, experiments, and examples you haven't thought about before—the tyrocidine operon? The impact of tetrodotoxin on the nervous system? The possibilities are practically endless.

Remember that all free-response questions seek to test your understanding and synthesis of biological theories and concepts, as well as your ability to interpret experimental data, not highly specific facts. Applying what you know to these unfamiliar scenarios will help you get a great score, even if you've never thought much about the particular experiment or scenario listed in the question. (The tyrocidine operon—it's still an operon and follows all the rules of other prokaryotic operon systems.) So don't panic when you see something unfamiliar. Realize that it will be related to something you do know and work from there.

# Chapter 3

# Natural Selection and Evolution

## Contributing Ideas to Darwin's Descent with Modification

In the 1700s and 1800s, the biological sciences were defined in terms of **natural theology** rather than scientific data and extrapolation. In fact, prior to any theories of evolution, the accepted belief was that organisms were unchanging. This belief was strongly held throughout the 18th century. However, increased exploration of the world generated observations that made it clear that organisms have changed over time. Explorers and scientists found that there were many more species of plants and animals than previously known. Many organisms that appeared to be different actually had similar body plans, and fossils of extinct organisms were found to be similar to organisms that were living.

**Charles Darwin** (1809–1882) built on the ideas of other scientists to develop his theory of "descent with modification" by natural selection. He was the first to propose a mechanism for evolution that was supported by evidence. Darwin proposed that changes in populations occurred when populations had differential survival and reproduction rates due to the presence of inherited variations.

While examining the fossil record, both Jean-Baptiste Lamarck and Charles Darwin agreed that species evolve over time, but each proposed a different mechanism:

- In the 19th century, Lamarck proposed the (incorrect) mechanism for evolution centered on acquired characteristics being inheritable. Lamarckism, as the idea came to be known, asserted that if a trait is used, it will be passed down to the next generation, but, if not used, then it will be discarded and not passed along. This theory of evolution was not supported by evidence; however, it was important because it proposed a mechanism for evolution that included the environment's importance in shaping changes in organisms.

> **DIDYOUKNOW?**
> Darwin did not use the term "survival of the fittest." It was Herbert Spencer, an English philosopher, biologist, sociologist, and political theorist, who coined this phrase in 1864.

- Darwin also recognized that species change over time, but he proposed a different mechanism for how that change occurs, which he called **natural selection**.

Darwin was influenced by the ideas proposed by a number of other scientific thinkers, as well as his own extensive observations of biogeography and of plant and animal breeding.

- Charles Lyell, a geologist, proposed that the Earth had been around for a long period of time, that geological processes—such as volcanic eruptions, that occur presently also occurred in the past (**uniformitarianism**), and that these types of processes, over a long period of time, account for large-scale changes in the Earth's physical characteristics (**gradualism**). These ideas led Darwin, and others, to conclude that the *strata* (and their fossils), observable in the exposed rock represent distinct time periods during the Earth's history.

- Thomas Malthus proposed that population size remains fairly steady, despite its capacity for exponential population growth, because of disease, wars, and limited resources. Darwin felt that this situation applied more generally to all species and further proposed that the availability of limited resources led to competition between members of a species.

# Natural Selection

By studying 12 different types of finches on the Galápagos Islands, Darwin made a link between the origin of a new species and the environment in which these species reside.

The **Theory of Natural Selection** shows that the reproductive success of an organism depends on its ability to adapt to the environment in which it resides. For example, several of the finches in the Galápagos Islands adapted their beak structure in order to find food.

## Postulates of Natural Selection

- If the environment cannot support the individuals who occupy it, then competition occurs between members of a species and affects the production of offspring.

- All members of a population have variations, and these variations are inherited (genetic). Some inherited variations within a population may enhance survival. Darwin called these variations **adaptations**. In other words, survival of individuals within a population will depend on their genetic background. Individuals with traits that promote survival will pass these traits to offspring, allowing them to be more "fit" for their environment.

- Over time, the fittest organism will survive, hence "survival of the fittest," and, therefore, changes in the population (genetic variation and mutations) cause variability and are an asset to a species. These population changes take place to benefit the reproduction of the population.

The result of natural selection is the adaption of population to their environment, thus giving them a competitive advantage to survive.

A genetic variation, such as the average beak length of a finch changing based on the season, is an example of adaptation. Such a trait manifests itself in order to provide an advantage in specific environmental conditions. For example, during the dry season, the average beak length gets slightly larger, giving the finches a better advantage to transverse terrain and outcompete other birds for seeds that are less abundant in a wet season. A larger beak indicates a competitive advantage and survival of the fittest.

### TEST TIP

For the AP Biology exam, it is almost certain you will need to know the observations Darwin made and the inferences he drew to develop his theory of evolution.

## Questions to Consider

1. A group of scientists living on the Galápagos Islands has been measuring the thickness of the beak of a particular seed-eating species of finch. During a prolonged period of drought, the average beak thickness increased. At the same time, the number of seeds the finches used for food decreased and the seed coats of the remaining seeds were thicker on average. What was being observed? What could you infer from these observations? (Hint: Look at Darwin's observations and inferences.)

2. Given four populations of organisms:

    Group 1 mates with Group 2
    Group 2 mates with Group 3
    Group 1 does not mate with Group 3
    Group 4 does not mate with any of the other groups

    How many species are present?

## Answers:

1. Members of a population have variations: birds have differing beak thicknesses and seed coats have various thicknesses. These traits are inherited. The prolonged drought reduced the food resources available to the birds. This selective pressure favored birds with thicker beaks that were more successful at utilizing resources. More birds with thicker beaks survived and produced offspring with this beak trait. Over time, this could result in a species of finches with, on average, thicker beaks. The seeds also experienced selective pressure for thicker seed coats. The seed coat provided protection from desiccation and/or predation. Over time, this resulted in selection for thick seed coats.

2. There are two separate species. Groups 1 and 2 interbreed and Groups 2 and 3 interbreed, so there is gene flow between Groups 1 and 3. Therefore, they are not genetically distinct. Group 4 does not interbreed with any of the other groups and does not exchange any genes with these groups; therefore, it is considered a distinct species.

# Chapter 4

# Evolution: An Ongoing Process

## Population Genetics and Hardy-Weinberg Equilibrium

**Population genetics** is the study of how alleles are inherited from generation to generation within a population.

**Population** is a group of individuals of the same species that share a gene pool, and it is the smallest unit in which evolution can occur. Examples of common traits that can be shared by individuals in a population can include four limbs and fur.

**Gene pool** is the total sum of genes within a population at a given time.

Darwin did not know the origin of these inherited variations in a population. However, now we know that these variations occur within an organism because of an organism's **genes**, so the environment—through natural selection—indirectly selects for or against genes by selecting for or against the phenotype (a.k.a. the variation or adaptation) that the gene produces. In this way, natural selection can change gene frequencies in a population.

Differences in members' DNA may produce different versions of a trait and thus different phenotypes within a population. The different versions of a trait are known as **alleles**. For example, a color trait of feathers may be expressed as the phenotypes' brown feathers or red feathers.

## Hardy-Weinberg Equilibrium

The Hardy-Weinberg Equilibrium indicates that the frequencies of two alleles do not change from generation to generation. A population is said to be in Hardy-Weinberg Equilibrium if the following five conditions are met:

1. It has a very large population sample.

2. There is no migration of individuals into or out of the population.

3. There is no mutation of either of the alleles.

4. There is no natural selection.

5. Random mating occurs between members of the population.

If gene frequencies do change significantly, at least one of these conditions is not being met in the population. Keep in mind that the Hardy-Weinberg principle of equilibrium is applicable to ideal situations and usually would not naturally occur in nature. However, this principle is very important in understanding population genetics.

Gene frequencies of a population are determined by using the following Hardy-Weinberg equation:

$$(p + q) = 1$$
$$p^2 + 2pq + q^2 = 1$$

$p$ = frequency of the homozygous, dominant allele (AA)
$q$ = frequency of the homozygous, recessive allele (aa)
$2pq$ = frequency of heterozygous allele (Aa)

The combined gene frequency must be 100%, so that $p + q = 1$.

## Sample Problem #1

Assume a population of 500 pea plants in which green is dominant to yellow. Use the chart below to see how to calculate the frequencies of all phenotypes.

Figure 4.1. Example Problem for Hardy-Weinberg Equilibrium

A = green, a = yellow

| Phenotype | Green | Green | Yellow |
|---|---|---|---|
| Genotype | AA | Aa | aa |
| Number of pea plants (total = 500) | 320 | 160 | 20 |
| Genotypic frequencies | 320/500 = 0.64 AA | 160/500 = 0.32 Aa | 20/500 = 0.04 aa |
| Number of alleles in gene pool | 320 × 2 = 640 A | 160 A + 160 a = 320 A & a | 20 × 2 = 40 aa  40 a |
| Allelic frequencies | 640 A + 160 A = 800 A  800/1000 = 0.8 A  p = frequency of A = 0.8 | | 160 a + 40 aa = 200 a  200/1000 = 0.2 a  q = frequency of a = 0.2 |

$$p^2 + 2pq + q^2 = 1$$
$p^2$ = frequency of AA = 0.8 × 0.8 = 0.64 = 64%
$2pq$ = frequency of Aa = 2 × 0.8 × 0.2 = 0.32 = 32%
$q^2$ = frequency of aa = 0.2 × 0.2 = 0.04 = 4%

$$p + q = 1$$
0.8 + 0.2 = 1 (Always check to make sure these numbers equal 1.)

## Sample Problem #2

Consider a large population of flowers in which 16% of the flowers are white (a recessive phenotype) and 84% of the flowers are red. What are the gene frequencies and the genotypic distribution in this population?

$q^2 = 0.16$
$q = \sqrt{0.16} = 0.4$
$p = 1 - 0.4 = 0.6$
$p^2 = 0.6 \times 0.6 = 0.36$      Homozygous dominant = 36%
$2pq = 2(0.6 \times 0.4) = 0.48$      Heterozygous = 48%

Keep in mind that in most populations, one or more of the Hardy-Weinberg conditions are not being met.

> **TEST TIP**
>
> Know the Hardy-Weinberg Equilibrium concept and how to use it. It is highly likely that there will be questions on the AP Biology exam that refer to it.

# Phenotypic Variation

**Microevolution** is the change in the frequencies of alleles or genotypes in a population from generation to generation (evolution on a small scale) and occurs if any of the five conditions of Hardy-Weinberg equilibrium are *not* met.

**Genetic drift** is defined as changes in the gene pool due to chance because of a small population. The small population directly contrasts the large population needed to maintain Hardy-Weinberg equilibrium. It causes a significant genetic change (microevolution) of a species if only a few members of a population migrate to found a new population.

It also causes genetic change (microevolution) anytime a species is reduced to very small numbers due to chance events, such as hurricanes, earthquakes, fires, or habitat destruction.

There are two types of genetic drift: the bottleneck effect and the founder effect.

The **bottleneck effect** is caused by changes in the gene pool due to some type of disaster or massive hunting that inhibits a portion of the population from reproducing. The small population directly contrasts the large population needed to maintain Hardy-Weinberg equilibrium. For example, northern elephant seals were hunted to near extinction. The surviving population has rebounded, but the genetic variation within this population is lower than in corresponding southern elephant seals.

The **founder effect** occurs when a few members of a population form a new colony. Because of the smaller sample size, the less the genetic makeup of the population. The small population directly contrasts with the large population needed to maintain Hardy-Weinberg equilibrium. For example, the Afrikaner population has a relatively high rate for the gene that codes for Huntington's disease. This is because the few founding colonists—by chance alone—had a high frequency of this gene, not because this gene was in particularly high frequency in the parent population (i.e., the population from which the colonists originally came).

**Gene flow** is the transfer of alleles from one population to another through migration. The gametes of fertile offspring mix within a population, providing genetic variation. Genetic variation directly contrasts the no gene-flow postulate needed to maintain Hardy-Weinberg equilibrium.

**Mutation** is a change in the genetic makeup of an organism at the DNA level. Mutation directly contrasts the no mutation postulate needed to maintain Hardy-Weinberg equilibrium.

**Nonrandom mating** is defined as individuals mating with those in close vicinity. Nonrandom mating directly contrasts with the random mating postulate needed to maintain Hardy-Weinberg equilibrium.

**Natural selection** is the reproductive success of organisms that is dependent on their ability to adapt to the environment in which they reside. Natural selection directly contrasts with the no natural selection postulate needed to maintain Hardy-Weinberg equilibrium. Keep in mind that any genetic variation with a population can increase that population's genetic diversity, even within the same species.

**Some phenotypic variations can significantly increase or decrease the fitness of an organism** and the overall population. Examples include DDT resistance in insects, the peppered moth, and sickle cell anemia. Humans can also impact other species through loss of genetic diversity within a crop species, overuse of antibiotics, and artificial selection.

Environments exert selective pressure on phenotypes so an organism's phenotype may increase or decrease its fitness. An example of selection for a phenotypic variation—or **microevolution**—is illustrated by the change in winter dormancy of a species of mosquitoes due to global warming.

The *Wyeomyia smithii* is a mosquito that survives the winter cold by going dormant when the days grow shorter. However, dormancy decreases the time the mosquitoes have for reproduction. Scientists gathering data on these mosquitoes have found that, on average, these mosquitoes are entering dormancy *later* than they did 30 years ago. Shorter, milder winters due to global warming exerted selective pressure on the mosquitoes to enter dormancy later, increasing time for reproduction. The allele that cued later dormancy has become more frequent in this population. This represents a shift in the allele frequency in the mosquito population and demonstrates that microevolution can occur over a relatively short period of time.

Natural selection can change a population in different ways. Four common types of selection are stabilizing, directional, diversifying, and sexual selection.

**Stabilizing selection** occurs when individuals in a population that have phenotypes that are intermediate along a spectrum of possibilities have an adaptive advantage. Some examples include the following:

- Human birth weight is affected by stabilizing selection because low-birth-weight and high-birth-weight infants are less likely to survive than those of intermediate weight.

- The number of eggs a bird lays is also affected by stabilizing selection because producing too few offspring means leaving less offspring to the next generation, but producing too many offspring requires additional care and feeding that may result in less healthy offspring.

**Directional selection** occurs when individuals in a population that have a phenotype at one end of the continuum have more of a chance of surviving than those who have intermediate phenotypes or those who have phenotypes at the other end of the continuum. Some examples include the following:

- If large size becomes an advantage in a changing environment for a mammal, large size will be favored over small or intermediate sizes causing a gradual shift in the species toward larger body size.

- If flower color varies among members of a species, and pollinators prefer the more brightly colored ones, brighter colored flowers will have a selective advantage, be pollinated more often, and a change in the flower color of the species to the brighter color will occur over time.

**Diversifying selection** (also called disruptive selection) occurs when the intermediate phenotype for a trait has a selective disadvantage. Some examples include the following:

- If insects have two different colorations that are both equally well camouflaged, but the intermediate color stands out against the background objects in the species habitat, two extreme phenotypes have a better chance of survival.

- If, in a population of frogs, those that are larger and those that are smaller have greater fitness than frogs of intermediate size, then the species may evolve to have two distinct sizes of individuals.

Figure 4.2. Results of Stabilizing, Directional, and Diversifying Selection

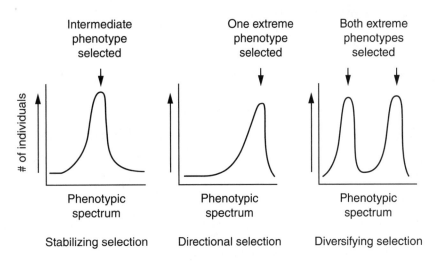

**Sexual selection** is a special type of natural selection in which a phenotype that is particularly effective in gaining a mate is preferentially selected, such as the showy feathers of a peacock or a behavioral adaptation that produces a better display of a mating behavior.

> **TEST TIP**
>
> If you are given a specific case of selection, be sure you are able to predict the direction of changes to allele frequency.

Artificial selection and genetic modification can also change a population. Humans can interfere with the natural processes of evolution through both artificial selection and genetic modification.

**Artificial selection** occurs when humans selectively mate organisms with traits felt to be desirable; this may increase or decrease genetic variations. Because of artificial selection, dogs have a greater number of genetic variations within their population than do wolves. Also, crop plants have decreased genetic variations because humans select specific crop plants for very particular growth and yield traits.

**Genetic modification** of food products involves the artificial introduction of genes into food crops, adding an artificial variation into a species. One potential risk of genetic modification of crops is unintended gene flow between the genetically modified crop and wild type plants. This gene flow could alter the genotype and phenotype of the wild type crops and make them less likely to survive, leading to extinction of the native plant species.

# Speciation

Throughout Earth's 4.5 billion-year history, new species have evolved by splitting off from existing species in a process called **speciation**. Although speciation is a constant biological process, its rate varies and it is often determined by changes in a population's surrounding ecosystem. For example, if a new habitat becomes available, then some organisms will spread into that area and likely evolve into a different species than the one left behind because of different environmental pressures.

Speciation occurs in the following ways:

In **allopatric speciation**, populations of organisms are separated by geographic isolation so that these two subpopulations that can no longer interact. A new species can then be formed following adaption to new surroundings. For example, if the climate changes and a large lake loses water, it may divide into smaller lakes. This would separate species into two (or more) geographic areas and prevent gene flow between the organisms, ultimately causing the two subpopulations to evolve into two distinct species.

**Adaptive radiation** is the evolution of a large number of species from a common ancestor. For example, the finches Darwin found on the Galápagos Islands are an example of adaptive radiation.

Once a new species evolves, **reproductive isolation** is necessary to keep the new species distinct from the old species. It consists of biological factors that act as barriers to reproduction. These factors are divided into prezygotic and postzygotic barriers.

**Prezygotic barriers** act to prevent either the mating process or fertilization if mating does occur. Isolating species by habitat (ground bird versus tree-dwelling bird), time of breeding season (spring versus summer), or behavior (different courtship rituals) will prevent different species from mating. If two species attempt a physical mating, then morphological differences, such as genital opening alignment, may prevent fertilization. Also, differences in gametes may prevent sperm from fertilizing an egg. For example, protein receptors on the egg of one species may not recognize the sperm from another species.

Occasionally a sperm and egg from two different species overcome the prezygotic barriers and form a hybrid zygote. Thus, **postzygotic barriers** exist to prevent this zygote from developing into a viable offspring or a fertile adult. For example, a mule, which results from a cross between a male donkey and a female horse. Although a mule is viable, it is sterile and unable to reproduce. Reproductive isolation is a distinct process from **geographic isolation**, which prevents reproduction by physically separating individuals from one another, greatly restricting gene flow between populations.

**Sympatric speciation** is a form of reproductive isolation that occurs in populations living in the same geographic area. However, some factor arises that prevents gene flow between subpopulations. An example occurs when a mistake during cell division creates a polyploid organism, which has an extra set of chromosomes. This is seen most often in plants when a diploid cell fails to properly divide and results in a tetraploid (four sets of chromosomes) offspring. The tetraploid offspring would not be able to reproduce with a diploid member of the species, resulting in the rapid evolution of a new species.

## Speciation Rates

There are two theories that explain speciation rates: gradualism and punctuated equilibrium.

**Gradualism**, proposed by Charles Darwin, states that changes in a population accumulate slowly and steadily over time. The differences in individuals occur so slowly that they are not noticeable over a short period of time (i.e., 1,000 years). However, over a longer period of time (thousands or millions of years), these changes become obvious.

**Punctuated equilibrium** was proposed in the 1970s by paleoanthropologists Niles Eldredge and Stephen Jay Gould. It posits that populations of organisms remain stable for long periods of time, then change quickly in spurts. DNA mutations often cause punctuated equilibrium. If a beneficial mutation arises and significantly impacts the survival of an organism having the mutation, it can quickly be found throughout the gene pool in just a few generations.

Drastic changes to an ecosystem can also cause punctuated equilibrium through extreme natural selection in which only a few organisms survive and their genes will dominate the population's future gene pool.

# TEST TIP

Natural selection acting on a population is the mechanism by which a species' characteristics change (evolve) over time. Remember to think about how one topic in biology relates to another.

## Questions to Consider

1. You are studying fossils found in sedimentary rocks. The rock you are studying consists of three distinct layers. In the bottom layer are many small creatures that appear to have approximately 10 legs and 5 body segments. The middle layer is thinner, made of black volcanic rock, and contains no fossils. The top layer has fossils that are similar in size and shape to those found in the bottom layer—they have 10 legs—but only appear to have 3 body segments. Explain what might have happened to cause these changes in the fossil record and determine if this is an example of gradualism or punctuated equilibrium.

2. When Darwin travelled to the Galápagos Islands, he observed a number of different finch species. He determined that each species was very similar except for their beaks, which were adapted to specific diets related to the food most abundant on the island inhabited by each species. What kind of speciation is this an example of and how did it occur?

3. The Colorado potato beetle eats the leaves of potato plants and causes hundreds of millions of dollars in damage each year to the agricultural industry. Since the 1860s, when chemicals were first used to kill the beetle, it has developed resistance to approximately 50 different chemical pesticides. Explain what happened to the Colorado potato beetle in terms of both genes and evolution. Why is this evidence that evolution is an ongoing process?

4. Indicate whether each of the following is likely to increase or decrease variations within a population:

    (a) Natural selection
    (b) Nonrandom mating
    (c) Immigration
    (d) Mutation
    (e) Recombination

5. Albinism is a recessive trait. In a certain population of 200 individuals, 18 people are albino.

   (a)   What are the genotypic frequencies within this population?
   (b)   What percentage of this population is NOT a carrier of albinism?

## Answers:

1. This is punctuated equilibrium. It is likely that a volcanic eruption caused this species to rapidly evolve.

2. This is an example of allopatric speciation. As finches moved onto new islands, they became geographically isolated from one another. The finches having beaks better adapted to eating that island's food source were more likely to survive and pass their beak genes on to their offspring.

3. Because of the variety in alleles and genes in a population, some beetles have gene mutations that provide resistance to pesticides. Before the use of pesticides, the gene pool would have had very few pesticide-resistant mutant genes in it. Over time, pesticide use has killed the beetles without resistant genes, causing the gene pool to shift towards the resistant gene. This is evidence of ongoing evolution because over the past 150 years, farmers have seen resistance to different chemicals develop.

4. (a)   Natural selection decreases variation because it is selecting for or against a particular variation, making one phenotype, and therefore genotype, more common than another.

   (b)   Nonrandom mating will decrease variation. This is a form of selection in which one mate's phenotype is favored over another, resulting in more mating opportunities for one kind of phenotype. Genes for the selected phenotype will, therefore, become more common. For example, if peacock females elect to mate with males that have more spectacular tail feathers, more matings will occur between females and spectacularly tailed males. The genes for spectacular tail feathers will increase in the subsequent generations. Nonspectacular genes will decrease.

   (c)   Immigration increases variation because individuals can bring in novel genes, thus variations, from other populations.

(d) Mutation increases variation because it is a change to DNA that potentially represents a new and different gene, thus potentially a new phenotype.

(e) Recombination increases the number of different combinations of genes, therefore, the number of different phenotypes that can occur.

5. (a) The trait for albinism is recessive, so its genotypic frequency is the same as its phenotypic frequency:

$$\text{Albino (aa)} = q^2 = 18/200 \text{ so}$$
$$q = \sqrt{18/200} = \sqrt{0.09} = 0.3$$

The dominant gene, A, is calculated:
$$A = p = 1 - q$$
$$= 1 - 0.3 = 0.7$$

So:

$$\text{Homozygous dominant (AA)} = p^2 = (0.7)^2 = 0.49$$
$$\text{Heterozygous (Aa)} = 2pq = 2 (0.7 \times 0.3) = 0.42$$
$$\text{Homozygous recessive} = q^2 = 0.09$$

(b) 49% of the population is NOT a carrier—that is, they do not have the recessive allele for albinism.

**Time for a quiz**
- Review strategies in Chapter 2
- Take Quiz 1 at the REA Study Center
  (www.rea.com/studycenter)

# Chapter 5

# Common Ancestry and Models of Evolutionary History

## Common Ancestry

A substantial body of evidence has accumulated in support of organismal change over time. The evidence for evolution comes from many scientific disciplines.

**Biogeography** is the study of organisms and how they relate to the environment. Some organisms may be unique in how they relate to the environment. Some organisms may be unique to certain geographies; hence, those organisms have adapted to live in that environment.

**Fossils** help indicate the progression of organisms from simple to complex. For example, transition fossils are fossils of animals that display a trait that helped the organism attain a competitive advantage. At one time, fossils show whales had limb-like appendages, indicating they may have been land dwellers.

**Comparative anatomy** is the study of anatomical similarities between organisms and provides information about how organisms are related and how selection has shaped the species.

- **Homologous structures** are structures with organisms that indicate a common ancestor. Organisms can appear to be very different but can have the same developmental origin. For example, a human arm, cat leg, whale flipper, and bat wing all have a similar structure, but different functions.

- **Analogous structures** are structures on different species of organisms that may resemble each other in appearance and/or function, but have different origins. Structures like these may arise as a result of convergent evolution and do not demonstrate common ancestry. For example, consider the eye of an octopus vs. the eye of vertebrates. They both function in vision, but closer inspection reveals they have very different developmental origins.

- **Vestigial organs** are remnants of structures that were at one time important for ancestral organisms. For example, the pelvic bones in snakes have no function for the snake, but provide evidence of an ancestral relationship with organisms that possess functional pelvic bones.

**Comparative embryology** compares the embryonic development of one organism to another.

**Molecular biology** is used in the study of evolution by looking at homology in DNA and protein sequences and genes; this study allows for an even broader level of comparison between organisms as different as prokaryotes, plants, and humans.

Organisms share conserved core processes, which signal their evolution from a common ancestor and how widely these processes have become among different species. For example, DNA and RNA are carriers of genetic information through transcription, translation, and replication. The genetic code of many organisms is shared and is evident in many modern living systems. Many metabolic pathways, like glycolysis, are conserved.

Structural evidence, such as cytoskeletons, membrane-bound organelles, linear chromosomes, and endomembrane systems, suggests that all eukaryotes are related.

The number of differences between organisms' DNA represents a *quantitative* measure of relatedness and overrides anatomical similarities or differences in determining relatedness. DNA mutation rates give information about how long ago a divergence between any two species may have occurred.

## Unity of Organismal Processes

Organisms are made up of cells that obtain and utilize energy and transmit genetic information from one generation to the next. These essential processes are often directed by genes that are identical—or nearly so—even when the organisms themselves are very different. These processes, then, are critical to an organism's survival, and any changes to these processes can harm an organism's ability to survive and reproduce.

Since critical processes are uniform in many organisms, these processes are considered to be **conserved**, and the genes that direct these processes accumulate mutations much more slowly—if at all—than do less critical genes. Conserved structures and processes often share a common function and origin, provide evidence for relationships among organisms, and provide evidence for a common ancestor for all living organisms. Whether or not a gene is conserved across different taxonomic groups can be determined by aligning genes or amino acid sequences of proteins that have the same or similar function among the groups.

Some structures and processes are common to all three domains of life (eukarya, eubacteria, and archaea). The following are examples of either cellular structures or processes that are shared by all living organisms:

- The phospholipid bilayer structure of the plasma membrane of cells is the same for all three domains of living organisms and has the same function, selective permeability.

- All domains of life are made up of a limited number of chemical building blocks that can be arranged in many ways. The twenty amino acids, for example, are common to all domains of life and make up an almost infinite variety of functional proteins.

- Whether an organism is single-celled or multicellular, the continuity of life depends on genetic information being passed from one generation to the next. DNA carries the genetic information that directs the activities and structure of a cell through RNA processing and the production of proteins. The genetic code, carried on DNA, is universal and the processes of replication, transcription, and translation are similar across all domains. In fact, it is possible to have a bacterium produce a eukaryotic protein because of the universality of the genetic code and production of proteins.

- Ribosomal RNA (rRNA) is critical to all life forms, thus the DNA that codes for rRNA is highly conserved. Comparisons of rRNA sequences provide evidence for relationships that are *ancient* because rRNA genes change very slowly. Homologies in rRNA can be detected across all three domains of life.

- Glycolysis and ATP production are processes that occur across all domains of life and the enzymes, proteins, and electron acceptors required for the processes are homologous.

In addition to common functions, the cell is the fundamental unit of life. The two broad categories of cells—**prokaryotic cells** and **eukaryotic cells**—have several basic features in common. However, the eukaryotic cell represents a significant evolutionary change from prokaryotes. Plants, animals, protistans, and fungi are characterized by eukaryotic cells.

The structural and functional similarity between the respiratory proteins of prokaryotes and eukaryotes gives support to the **theory of endosymbiosis** as the origin of eukaryotes. That is, organelles such as mitochondria and plastids originated as ingested prey that was not digested but, instead, formed a mutual relationship with its host. If a cell—prokaryote or eukaryote—undergoes aerobic respiration, the proteins necessary for electron transfer are homologous, supporting a common evolutionary ancestor. In the eukaryote, these proteins are found on the internal membranes of the mitochondria and plastids, but in prokaryotes they are found in the plasma membrane.

## Evolution Continues to Occur

Scientific evidence supports the premise that evolution continues to occur. Some examples include:

- Emergent diseases

- Phenotypic change in a population (such as Darwin's finches in the Galápagos)

- Chemical resistances caused by mutations. Recently, bacterial resistance to antibiotics, insect resistance to pesticides, and weed resistance to herbicides also have provided support that evolution is an ongoing and constant process.

*Example:* The evolution of bacteria having gene mutations that confer resistance to antibiotics has occurred rapidly since the onset of widespread use of antibiotics. When random mutations arise in a cell, giving this cell resistance to an antibiotic, all offspring also have the resistant mutation. Thus, bacterial colonies, consisting of thousands of individual bacterial cells, will have some cells easily susceptible to antibiotics, while others are resistant. Ultimately, our use of antibiotics kills the weaker strains and often leaves behind the resistant

**DIDYOUKNOW?**
About 15% of human genes are like bacterial genes, 25% are like single-celled fungi, 50% are like fruit flies, and 70% like frogs.

strains, thus shifting the gene pool of the bacteria towards the resistant alleles and causing evolution of those bacteria.

Another example supporting ongoing evolution are the changes found in heart chambers of vertebrate animals. Most vertebrates use a system of double circulation that separates oxygenated and deoxygenated blood in a heart having multiple chambers. Amphibians and reptiles use a three-chambered heart, while birds and mammals use a four-chambered heart. All these animals have two atrial chambers. However, the ventricle chambers change from undivided in amphibians to partially divided in reptiles to completely divided in mammals and birds.

What makes organisms different (or alike) are their heritable variations. The origins of many evolutionary variations are changes in important developmental steps. It is hypothesized that most evolutionary innovations result from changes in regulatory genes operating on conserved core processes.

Regulatory genes called **homeotic genes** determine the major features of organisms. One set of homeotic genes, the *Hox* genes, determine where anatomical structures will develop on an organism's body and what form they will take. A small change to *Hox* genes, then, can effect a major structural change to a species. For example, humans and chimpanzees have very similar skull structures as infants, but changes in growth rates between the two species produce profound differences in the shapes of their skulls. As a result, the chimpanzee has an elongated skull with a stronger jaw when compared to a human.

# Models of Evolutionary History

**Phylogeny** is the field of science that deals with understanding and identifying the evolutionary relationships among the diversity of life on Earth. Scientists use fossil evidence, phenotypic similarity, mating compatibility, and molecular comparisons to establish lines of descent.

**Phylogenetic trees and cladograms** attempt to model evolutionary relationships and the chronological sequence of branching within a group of organisms. They represent traits that are either derived or lost due to evolution, such as opposable thumbs, the absence of legs in some sea animals, and the number of heart chambers in animals. They illustrate that speciation has occurred and when two groups were derived from a common ancestor. Phylogenetic trees and cladograms can be constructed from either morphological similarities or from DNA and protein sequence similarities by utilizing a computer program that can measure the organisms' interrelatedness. Each provides a dynamic snapshot that is constantly being revised.

A **phylogenic tree** is a diagram depicting the evolutionary relationship between groups of organisms, and is a working hypothesis.

Figure 5-1. Phylogenic Tree: Plant Evolution

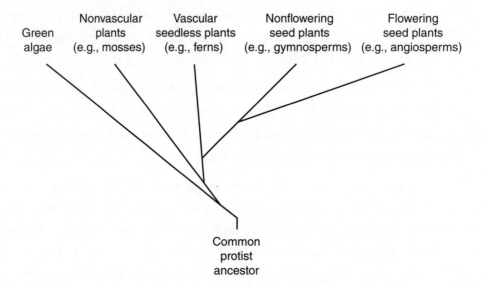

**Cladograms** are a type of evolutionary tree diagram that groups organisms based on shared **derived traits**. A derived trait is a novel feature that has evolved and is shared by the descendants of the ancestors from which it evolved.

Figure 5-2. Cladogram

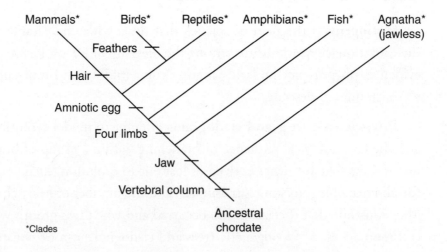

More and more, molecular data showing variations in DNA and subsequent proteins among different organisms are used to establish phylogenetic trees because closer evolutionary relationships are indicated by increased molecular homology. Comparisons of organisms using molecular data require computer programs that can statistically analyze sequences to determine how alike the sequences are. For example, a small change in DNA, such as a deletion, can make a highly conserved sequence look more different than it actually is in reality.

Remember that phylogenetic trees and cladograms are *hypothesized* relationships that can be falsified and changed as a result of new information.

**Taxonomy** is the science of classifying and naming organisms. Organisms are classified into hierarchical groups based on their similarities, and taxonomy helps to efficiently organize information about living things.

Currently, the taxonomic groups, from broadest to most specific, are as follows: *kingdom, phylum, class, order, family, genus,* and *species*.

Figure 5-3. Hierarchical Classification System

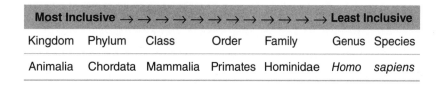

Naming an organism is based on Carolus Linnaeus' **binomial nomenclature**, which assigns a two-word name to each organism once the organism's species has been identified. For example, *canis lupus* is the scientific name given to the gray wolf. *Canis* is the genus name and *lupus* is the species name. Within its genus, a species name is unique.

# TEST TIP

For the AP Biology exam, be sure you are able to apply *binomial nomenclature* to determine the relative closeness of two organisms.

## Questions to Consider

1. Why is it possible to use an organism as different as yeast to study the actions of a drug to be used in humans?

2. The cdc2 (cell division cycle 2) gene is found in every eukaryotic organism and functions as an activator of mitosis. In fact, the human cdc2 gene can be substituted into cdc2-deficient yeast cells, and the yeast cells will undergo cell division correctly. What can be inferred about this nearly universal eukaryotic gene?

## Answers:

1. Up to 25% of the genes found in humans are homologous to those found in single-celled fungi such as yeast. If the drug to be tested acts on the products of that shared set of genes, the yeast response to the drug can possibly shed light on what the human response to the drug might be.

2. If a gene is homologous across vastly different species—the human cdc2 gene was sufficiently close to the yeast cdc2 gene to actually work effectively in the yeast—then the gene has been highly conserved and must be ancient. Furthermore, the gene must be part of a process that is critical to an organism's survival. Few changes in a gene means that any mutation in the gene has been heavily selected against and thus not maintained in the gene pool. This homology is also evidence for lines of descent from common ancestry among eukaryotes.

# Chapter 6

# The Origin of Life

## Origin of Life—Hypothesis and Evidence

Even the simplest living organism must have the following minimum requirements to be a self-perpetuating entity subject to developing increasingly complex adaptations through natural selection:

- Availability of **organic nutrients** and/or the availability to make organic molecules from inorganic molecules

- A **barrier** to delineate itself from the environment

- A method of **passing its characteristics on to its offspring**; that is, a genetic material that specifies its own characteristics

Evolution through natural selection explains why life on Earth is so diverse and complex. However, it doesn't tell us how the first cells came into being. Although we cannot go back in time to find Earth's early matter, through experimental research, scientists have developed several hypotheses about the origin of life.

Organic compounds can be produced from inorganic compounds. In the 1920s, the scientists Alexander Oparin and John Haldane independently suggested that organic molecules could have been synthesized on primitive Earth through nonbiological means. Essentially, the Oparin-Haldane hypothesis proposed that inorganic molecules in the atmosphere rearranged into organic molecules (life's building blocks) using energy provided by either lightning or UV radiation.

Primitive Earth was thought to have the following atmospheric molecules—water ($H_2O$), methane ($CH_4$), hydrogen ($H_2$), and ammonia ($NH_3$)—and no oxygen. These inorganic precursors of organic molecules on primitive Earth could have been formed as a result of an electric spark and the lack of oxygen. Crude organic molecules including sugars, lipids, amino acids, and nucleic acids were formed.

In 1953, two scientists at the University of Chicago, Stanley L. Miller and Harold C. Urey, tested the Oparin-Haldane hypothesis by simulating primitive oceanic and atmospheric conditions. They placed ammonia gas, methane, water vapor, and hydrogen gas into a flask, then added an electric current to the system to simulate lightning. This experiment produced several simple organic molecules including amino acids, chains of hydrocarbons, and formaldehyde. Miller and Urey's experiment provided evidence that organic molecules can be synthesized from inorganic molecules without using living organisms.

Organisms first arose without contact with the atmosphere. Organisms can live under extreme conditions such as in a solution or on solid, reactive surfaces. Organic monomers have been found in the ocean. Membrane-bound vesicles will spontaneously form when lipids are placed in water due to the hydrophobic nature of lipids. These vesicles would have been primitive protocells containing polymers such as polypeptides, sugars, and nucleic acids.

Other experiments have used **solid surfaces**, such as hot rock, to replicate conditions on early Earth. The results have provided evidence that organic monomers spontaneously form complex polymers without the assistance of enzymes.

Organic compounds can spontaneously form and maintain internal environments. It was found that organic compounds can spontaneously form spheres (*protobionts*, called *coacervates*, *microspheres*, and *liposomes*) that can maintain a unique internal environment compared with their surroundings.

RNA and other organic molecules are capable of self-replication, and both can act as genotype and phenotype. The RNA World Hypothesis states that RNA was the first genetic material, not DNA. RNA molecules spontaneously fold into specific shapes (phenotypes) and can catalyze their own synthesis (spontaneously reproduce). In other words, not only does RNA help make proteins, it can also act as a biological catalyst.

RNA molecules with enzymatic activity (ribosomes) can catalyze chemical reactions and make copies of RNA. Thus, they provide a way for protocells to both synthesize more genetic material (being able to replicate is a key characteristic of life) and a way to facilitate biological reactions (processing energy is another characteristic of life).

## Earth's History and the Origin of Life

There is geographic evidence that supports that a common ancestry and that evolution is and has been occurring. The Earth is most likely around 5 billion years old, and it is believed that the Earth's environment was too hostile for life until about 3.9 billion years ago.

Evidence of the origin of life on Earth is found through the earliest fossil records that date back 3.5 billion years ago. In fact, fossils continue to be evidence that supports speciation and extinction as well. Scientists examine changes in fossils found in sedimentary rock, called the stratum, to establish an evolutionary history of a species or a group of organisms. Radioactive dating is used to determine the age of a particular fossil.

There is also molecular and genetic evidence to support that all living things come from a common ancestor. The origin of life is believed to have started with the **anaerobic prokaryotes** that emerged approximately 4 billion years ago, and the earliest living organisms were unicellular and had a genetic code and the ability to evolve and reproduce.

About 2.5 billion years ago, **prokaryotes** diverged into two types—bacteria and archaea. Oxygen accumulated in the atmosphere as a result of photosynthetic bacteria, and **eukaryotes** emerged about 2 billion years ago via the Endosymbiotic Theory.

Prior to 500 million years ago, life was confined to aquatic environments. Plants eventually found a foothold (root system) on Earth via a symbiotic relationship with fungi.

Ultimately, the fossil and genetic records support the statement that evolution has occurred in all species on Earth. The origin of mammals provides a specific example. Mammalian evolution took approximately 120 million years and was traced by examining morphological differences and similarities in jawbones of both extinct and extant tetrapods.

> **DIDYOUKNOW?**
> Stromatolites are commonly found in lakes and lagoons with excessively high salinity. One great source of them is Shark Bay in Western Australia. Recently, a new chlorophyll, f, was discovered in stromatolites in the bay.

Genetic and biochemical evidence indicates that all existing organisms share a common ancestor. All cells use the same genetic code. This means not only do all living organisms use the nucleotides A, T, C, G, and U, but the **codon** combinations code for the same amino acid. For example, the codon AUG codes for the amino acid methionine in prokaryotes and eukaryotes.

All cells also use the same molecular building blocks of nucleic acids, carbohydrates, proteins, and lipids.

## Extinction and Adaptive Radiation

In addition to the appearance of new species on Earth, many species have also gone extinct.

**Extinction** occurs when a species of organism is permanently lost—meaning that all organisms of that species are not living. A certain amount of background extinction occurs regularly and naturally. Extinction rates become rapid during ecological stress, leading to a mass extinction over a short period of time.

There have been five major mass extinction events throughout Earth's history in which at least 50% of all species died out. For example, during the *Cretaceous extinction*, which occurred approximately 65 million years ago, about 50% of species, including almost all of the dinosaurs, became extinct.

### TEST TIP

Knowing the names and dates of the five mass extinctions is not required for the AP exam; however, be prepared to use data to determine that extinction has occurred.

The rapid development of a new species from a common ancestor may occur after a significant genetic change in a member of a species, or after a new habitat becomes available due to extinction of another (or many) species. **Adaptive radiation** causes an increase in speciation. It occurs after mass extinctions. For example, mammalian evolution after the Cretaceous period provides an example of how rapid speciation can occur through the process of adaptive radiation. Fossil evidence from the Cretaceous period supports that mammals were small and not diverse because they were outcompeted by dinosaurs. However, the extinction of dinosaurs allowed mammals to evolve and diversify into many large and successful land species.

## Questions to Consider

1. If there had been no ammonia in Earth's primitive atmosphere, how might that have changed the building blocks of life?

2. Why would membranes spontaneously form in water?

3. You find a new species of bacteria and sequence its DNA. You discover that the codon AUG codes for lysine instead of methionine. What significance would this discovery have on the field of biology?

## Answers:

1. Proteins (amino acids) and nucleic acids (nucleotides) would not be formed; but carbohydrates and lipids could be formed.

2. Lipids are nonpolar hydrophobic due to their long hydrocarbon chains, so they will be attracted to one another in order to stay away from polar water molecules.

3. Because life shares a common genetic code, finding an organism that does not fit with this code implies that it arose from an ancestor different from other life on Earth.

# AP Biology Labs: Evolution

## Chapter 7

## Lab 1: Artificial Selection

### Concepts

This lab deals with evolution and natural selection, more specifically with **artificial selection** in Wisconsin Fast Plants®. Artificial selection is choosing which organisms to breed in order to increase the likelihood of desirable traits. The question you answered in this lab was: Did artificial selection change the genetic makeup of your population of plants? When studying evolution, you must realize that all populations show phenotypic variation (see **phenotype**), or differences in physical traits. The goal of this lab was to choose one phenotypic extreme and to breed plants having that trait. By picking which plants to cross in this lab, your class influenced the outcome of the offspring.

In this lab, it is likely that your teacher had you count the number of trichomes on each Fast Plant. Trichomes are small hairs found on the stems and leaves. Your class raised many Fast Plants and chose a sample of the population that had the most trichomes. By crossing the "hairiest" plants, you probably caused **directional selection**, which causes an increase in a trait in one direction (for example, more plants displayed trichomes).

## Science Practices

One method you might have used in this lab is sampling. Sampling is when you count a small portion of a population and then extrapolate to the entire population. For example, a plant that had many trichomes probably had so many that it would be difficult to count them all. In this method, you assume that the rest of the plant would express a similar density of trichomes per unit area. Thus, you could sample by counting the number of trichomes on the first leaf closest to the soil on all plants and compare the number of trichomes found on all plants. In this lab you selected two parents to breed. Then you cross-pollinated the two plants by taking the pollen (containing the male gamete) of one plant and spreading it to the flower (containing the female gamete) of another plant.

## Data Analysis

Most of the data collected in this lab can be analyzed with descriptive statistics, which are useful for identifying trends in data. The table below (Table 7.1) describes these statistics, how they are calculated, and the type of data that would indicate that your cross-pollination was successful in causing directional selection of plants with more trichomes.

Table 7.1.   Statistical Methods

| Statistical Method | How to Calculate | Data Supporting an Increase in Trichomes |
|---|---|---|
| Mean | Take the average of all data. | An increase in the mean number of trichomes on all plants in the offspring as compared to the parents:<br><br>Mean = 4 trichomes in the parents<br><br>Mean = 8 trichomes in the offspring |
| Median | Divide the distribution into halves. Half the plants had more trichomes; half the plants had less trichomes. | An increase in the median in the offspring:<br><br>Median = 1 in the parents<br><br>Median = 7 in the offspring |
| Mode | Take the measurement that occurs most often. | An increase in the mode of the offspring:<br><br>Mode = 0 in the parents<br><br>Mode = 16 in the offspring |

One of the graphs that can be produced from data in this lab is a histogram to demonstrate frequency distribution of a particular trait. In creating a histogram, the data is numerical and continuous. The *x*-axis records the number of times the trait was expressed (in this case, the number of trichomes on a plant); while the *y*-axis shows the number of times a measurement or value was obtained. Examples of two possible histograms created with data from this lab are found in Figure 7-1.

**Figure 7-1.** Distribution of Trichomes in Two Generations of Fast Plants

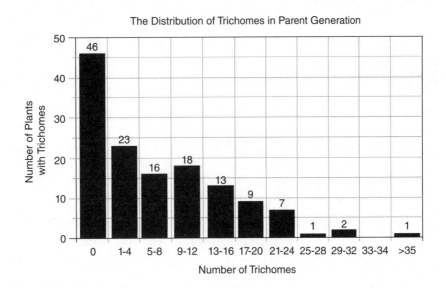

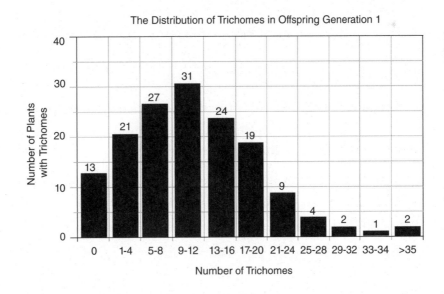

*Source: http://nces.ed.gov/nceskids/index.asp*

## Conclusions

It is likely that your class saw an increase in trichome numbers on your offspring plants. For example, in the histograms above showing frequency distribution of trichomes, it is clear that the offspring generation of plants expressed more trichomes. By choosing to cross the plants with the most trichomes, you increased the likelihood that their offspring would have many trichomes. Thus, artificial selection changed the genetic makeup of your plants, and the gene pool of the offspring has a higher number of trichome-bearing alleles.

# Lab 2: Mathematical Modeling—Using Hardy-Weinberg

## Concepts

There is a good chance that you might not have completed this AP lab due to constraints such as computer or spreadsheet availability. However, the concept of Hardy-Weinberg equilibrium is one of the keystones of evolutionary biology, so it would be wise to review the concepts for this lab. Essentially, Hardy-Weinberg (H-W) calculations are used to mathematically model the rate of evolution in a population. Populations of organisms consist of **gene pools**, or all the alleles in that population. It is the gene pool that evolves over time because the frequency of alleles in a population fluctuates due to pressures such as natural selection, migration of individuals, and random mutations.

It is important to remember that the H-W equation is actually a null hypothesis. It assumes the absence of any evolutionary factors. Essentially, if allele frequencies (as calculated with the H-W equation) do not change over time, then no evolution is occurring. However, if allele frequencies do change from generation to generation, we assume that evolution is occurring in a population.

H-W is difficult to simulate with living organisms in a classroom because populations consist of many more individuals than can be managed in a classroom or laboratory setting, and few organisms are suitable to study over multiple generations. Thus, this lab utilizes mathematical modeling using a computer spreadsheet to determine how a hypothetical (and infinitely large) gene pool evolves from generation to generation.

## Science Practices

In this lab, you used spreadsheet software to create a gene pool for a population and to measure any changes in the gene pool over time (evolution). The lab was broken into four parts. Part one formulated the guiding question of the lab: how do allele frequencies change in a population?

Part two established the use of one hypothetical gene in a population of diploid organisms. This gene had two alleles: A and B (two different letters were used to make it easier to read in the spreadsheet—in genetics problems we would have chosen A and a). Part two also assumed that the population had an infinite gene pool (this is why a model is useful) and that gamete selection was random.

Part three described the life cycle of a population of our hypothetical organism. In this population, all gametes had an equal chance of being used in fertilization, and all zygotes survived into adulthood. Additionally, no organisms left or entered the population and no mutations occurred. (You might not have realized it at the time, but the conditions established in parts two and three are the same ones that the scientists Hardy and Weinberg described as preventing a population from evolving.)

Finally, in part four you created and manipulated data in a spreadsheet to quantitatively describe the population over successive generations.

## Conclusions

We're going to skip the details of the spreadsheet because it's not relevant for this review. However, as you used the spreadsheet and manipulated the models, you should have established some basic rules of allele frequencies and organisms in the population.

1. The probability of two gametes containing A alleles creating an organism in the next generation is $p^2$.

2. The probability of two gametes containing B alleles creating an organism in the next generation is $q^2$.

3. The probability of A and B gametes or B and A gametes creating an organism in the next generation is $2pq$.

Thus, through this mathematical simulation, you established the mathematical rule of the Hardy-Weinberg equation that predicts genotypes in the next generation as $p^2 + 2pq + q^2 = 1$.

# Lab 3: Comparing DNA Sequences—BLAST Lab

## Concepts

Historically, scientists examining evolutionary relationships among organisms used characteristics such as morphology (physical structure), physiology, and fossil records to compare individual species. More recently, gene sequences have been used to determine relatedness. Many human genes are similar or the same as those found in other species—the more genes we have in common with an organism, the more related we are through evolutionary history. For example, humans share about 99% of our DNA with bonobos and chimpanzees, telling us that we are closely related to these species.

Once species' evolutionary relatedness is established, a cladogram (or phylogenetic tree) is constructed. This is a pictorial representation showing relationships. A simple cladogram is shown in Figure 7-2. This cladogram was constructed using shared derived characteristics, in this case particular physical characters such as hair. The closer two species are on a cladogram, the more derived characteristics they share, the more related they are, and the more recently they had a common evolutionary ancestor. In Figure 7-2, the human and mouse (both mammals) shared a common ancestor more recently than the human and the fish. Cladograms also can be constructed using percent similarity in genes. For example, if a human and a chimpanzee gene share 98.9% similarity and a human and a dog share only 92% similarity of that gene, we would place the human and chimpanzee closer together on the cladogram.

**Figure 7-2. A Simple Cladogram of Animal Species**

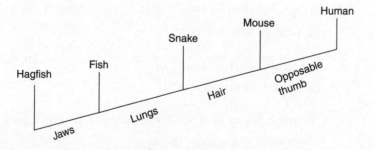

## Science Practices

You began this lab by examining a fossil animal for morphological similarities to animals found on a detailed cladogram. You probably noticed that it had vertebrae,

4 legs, and a tail, and hypothesized that it is related to the crocodilians or birds. Then you downloaded three gene files from DNA isolated from soft tissue of this fossil.

To analyze the gene sequence, a bioinformatics (a field combining statistics, modeling, and computer science) tool called BLAST (Basic Local Alignment Search Tool) was used. BLAST is a library of genomes for many organisms. It can be used to compare and determine similarities between multiple gene sequences. After you made a hypothesis, you uploaded your gene files to determine what organisms the gene sequences matched.

Finally, you used BLAST to research a gene you were interested in, such as catalase, insulin, or actin. In this research you might have found organisms with similar gene sequences to humans or discovered if there were genes found only in humans.

## Data Analysis

How to use BLAST is irrelevant to the lab, but the data that it returned is important. The BLAST results returned numbers that you used to establish if the data supported your hypothesis. The higher the score (or lower the e value) returned to you, the more similar the genes are. For example, if the BLAST analysis reported one of your gene sequences was highly similar to the sequence of a chicken, it supported your initial hypothesis that the fossil organism fits on the cladogram near the crocodilians and birds.

## Conclusion

Although there are many ways scientists can establish evolutionary relatedness, comparing gene sequences is a reliable tool when DNA evidence is available. Evidence from genomics can affirm morphological or physiological evidence, strengthening our understanding of the evolution of individual species on Earth.

**Time for a quiz**
- Review strategies in Chapter 2
- Take Quiz 2 at the REA Study Center
  (www.rea.com/studycenter)

# Chapter 8

# Energy and Matter

## Energy Changes—The Chemistry of Life

**Energy** is the capacity to do work. Energy is fundamental to life and is required to create and maintain living organisms. Organisms harness energy in different ways, but they are all bound by the laws of thermodynamics.

The **1st Law of Thermodynamics** state that energy can neither be created nor destroyed, but can change from one form to another and be transformed. For example, plants convert light energy from the sun to make glucose, a form of chemical energy.

The **2nd Law of Thermodynamics** states that every energy transfer increases entropy of the universe (disorder). No living system will violate the 2nd Law of Thermodynamics.

In other words, the energy transformations that take place are not 100% efficient, so, with every transformation, there is energy that is "lost" to an unusable form, usually in the form of heat. Since living systems follow the Laws of Thermodynamics, that is, energy is constant throughout the universe, energy can never be completely lost but instead is just converted to heat. Yet, without energy input, systems become disordered—that is, their **entropy** increases, and since cellular metabolic pathways are an orderly, enzyme-mediated series of reactions, energy input is often needed.

**Free energy** is the energy available in a system to do work. Organisms need this free energy to maintain organization, to grow, and to reproduce. Within a cell system, the free energy of a reaction, represented by $\Delta G$, determines whether the reaction is exergonic or endergonic.

**Exergonic reactions** release free energy. Therefore, the free energy of the products is lower than the free energy of the reactants. This means that there is a release of energy ($-\Delta G$). Furthermore, exergonic reactions are spontaneous.

$$AB \rightarrow A + B + \text{Energy}$$

In **catabolic reactions**, reactant(s) are broken down to produce product(s) containing less energy. The energy released can be used for reactions that require energy.

**Endergonic reactions** require free energy. Therefore, the free energy of the products is greater than the free energy of the reactants. This means an input of energy ($+\Delta G$) is necessary for the reaction to occur.

$$A + B + \text{Energy} \rightarrow AB$$

**Anabolic reactions**, those required for building large molecules from smaller molecules, are usually endergonic. The reactant(s) in this type of reaction are joined together to produce product(s) containing more energy. The free energy required by anabolic reactions is often provided by ATP produced in catabolic reactions.

Figure 8-1. Exergonic and Endergonic Reactions

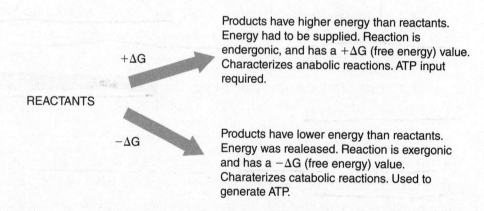

*Adenosine triphosphate (ATP)* carries energy in its high-energy phosphate bonds. The structure of ATP (see Figure 8-2) has three negative phosphate groups bonded adjacent to each other. The triphosphates form a relatively unstable molecule because the last phosphate group is easily removed from ATP and transferred to a reactant.

### Figure 8-2. Chemical Structure of ATP

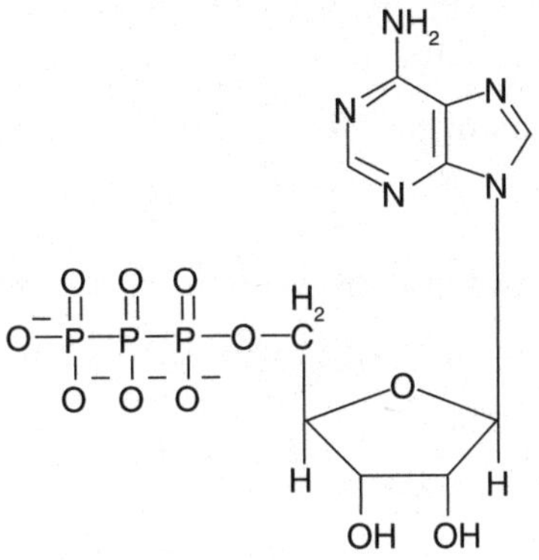

ATP is formed from adenosine diphosphate (ADP) and inorganic phosphate.

$$ADP + P_i + Energy \rightarrow ATP$$

Conversely, when ATP is broken down into ADP and $P_i$ via hydrolysis, energy is released (*exergonic*) that can be used in *endergonic* reactions. This removal of the phosphate group yields about −7.3 kcal/mol and its addition to a reactant destabilizes it, making the reactant more reactive.

### Figure 8-3. The Breakdown of ATP

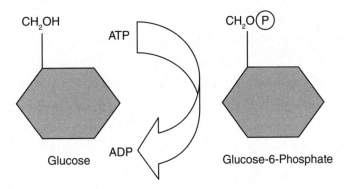

In addition, ATP can donate one of its phosphate groups to a molecule, such as a substrate or a protein, to energize it or cause it to change its shape.

Living systems require a consistent input of free energy and an ordered system. This free energy input allows for a system's order to be maintained. If either order in the system or free energy flow were to occur, death can result. Biological processes help to offset increased disorder and entropy and to help maintain order within a system. Thus, energy input into the system must exceed the loss of free energy in order to maintain order and to power cellular processes. Energy storage and growth can result from excess acquired free energy beyond the required energy necessary for maintenance and order within a system. Changes in free energy can affect population size and cause disruptions to an ecosystem.

Cells engage in chemical work, and this work involves activities such as molecular movement, synthesis of large molecules from smaller ones, cell reproduction, and the establishment of chemical gradients. The sum total of all the chemical work that goes on within a cell is called **metabolism**, which requires a constant input of energy.

The energy that enters a biological system takes a one-way path through the system. It enters the system as a **potential energy** form, such as in the chemical bonds of glucose, and is transformed within the cell to a usable **kinetic** form of energy. ATP is the main usable energy carrier in biological systems and is used to accomplish metabolic work.

Reproduction and the rearing of offspring require free energy beyond what is normally required for the maintenance and growth of an organism. Available energy can vary, and different organisms utilize a variety of reproductive strategies as a consequence.

**Asexual reproduction** requires less energy output by the parent, but results in offspring that are genetically identical to their parent. The lower energy input can result in greater reproductive rates. The genetic homogeneity means that asexual reproduction limits the potential environments available to the offspring.

**Sexual reproduction** requires formation and delivery of gametes, finding a mate, and providing food for the developing young. Some sexually reproducing animals, such as insects, put their reproductive energy into producing many offspring with little to no parental care given to the young after birth. Some organisms produce few offspring, but invest a lot of parental care after birth. Both means of sexual reproduction are strategies that have survival advantages and both require energy trade-offs.

**Seasonal reproduction** is an example of a reproductive strategy that maximizes the energy needed to produce offspring. The **life-history strategy** is another example that is utilized by biennial plants and reproductive diapause, for example.

Organisms utilize free energy to help regulate their body temperatures and metabolisms. The mechanisms through which these are done include the following:

- **Endothermy** is the use of internal thermal energy that is generated by metabolism to maintain an organism's body temperature. Endotherms (warm-blooded animals) maintain a fairly constant body temperature, which allows them to be active in a wide variety of environments. But, the downside to constantly maintaining body temperature is that their energy requirements will be much higher than an equivalent-sized ectotherm.

- **Ectothermy** is the use of external thermal energy to assist in the regulation of an organism's body temperature. An ectotherm relies on the environment to supply heat energy and, so, an ectotherm regulates its body temperature by behavior. For example, a lizard will bask in the sun to warm its core temperature. Ectotherms are more limited in the range of environment they can tolerate, but their energy requirements are relatively low in comparison to endotherms.

The size of an organism can affect its thermoregulation and subsequent metabolic needs. In other words, smaller organisms generally have higher metabolic rates. For example, small endotherms, such as shrews and mice, have a relatively high surface area-to-volume ratio. This means that they lose body heat relatively more quickly than a large endotherm. Thus, there is an inverse relationship between the surface area to volume ratio in a small versus a large animal, and the caloric needs of a small endotherm will be relatively greater than those of a large endotherm. For example, a shrew consumes about 900 mm$^3$ $O_2$/g of body weight/hr, whereas an elephant consumes about 100 mm$^3$ $O_2$/g of body weight/hr. The same correlation of size to metabolic needs is not seen in ectotherms.

## Energy Coupling

A **coupled reaction** is a chemical reaction having a common intermediate in which energy is transferred from one reaction to another. A system can maintain order by utilizing coupling cellular processes that increase entropy (causing negative changes in free energy) with those that decrease entropy (causing positive changes in free energy).

The molecule that is essential for coupling reactions and cellular work is **ATP** (adenosine triphosphate). In biological systems, endergonic reactions have to be coupled with exergonic reactions in order to occur. The source of the energy for biological systems is chemical bond energy of food molecules that is converted to ATP. ATP is produced by using the energy that is released when electrons are moved from a high-energy position (food molecules) to a lower energy position (usually water).

Exergonic reactions, like ATP → ADP, provide coupling energy for both endergonic and exergonic reactions. ATP, then, through the transfer of a phosphate group in a process called **phosphorylation**, is an example of an energetically favorable reaction because it allows for a negative change in free energy. This free energy can then be used to maintain or to increase order within a system that is coupled by reactions that demonstrate changes in positive free energy.

The processes of cellular respiration and photosynthesis are coupled to each other. The products of one reaction end up being the reactants in the other.

Figure 8-4. Coupling of Cellular Respiration and Photosynthesis

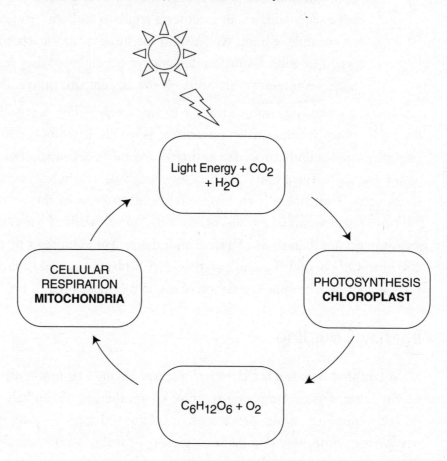

Electron transport and oxidative phosphorylation are also examples of coupled reactions.

# Modes of Energy Capture

All organisms, including plants, require a source of chemical energy for nutritional use in their biological systems. The following are different types of ways that organisms capture this free energy:

- **Autotrophs**, also referred to as "self-feeders," create their own organic molecules or food. This group of organisms is also known as **producers**.

- **Heterotrophs** cannot create their own organic molecules or food. This group of organisms is also known as **consumers**. One process that allows heterotrophs to acquire nutrition is hydrolysis, which helps them metabolize carbohydrates, proteins, and lipids as sources of free energy.

Table 8-1.  Modes of Nutrition

| Mode of Nutrition | Description; Examples (Other Nonprokaryote Examples) |
|---|---|
| Photoautotrophy | Use light as an energy source and gain carbon from $CO_2$; cyanobacteria (also plants and some protists) |
| Chemoautotrophy | Use an inorganic energy source and gain carbon from $CO_2$; some archaebacteria |
| Photoheterotrophy | Use light as an energy source and gain carbon from organic sources; some prokaryotes |
| Chemoheterotrophy | Use an organic energy source and gain carbon from organic sources; most prokaryotes (also animals, fungi, and some protists) |

Biological systems can capture energy at multiple points in their energy-related pathways. Some examples of these pathways include the Krebs cycle, glycolysis, the Calvin cycle, and fermentation. Energy capturing processes, such as $NADP^+$ in photosynthesis and oxygen in cellular respiration, use different types of electron acceptors.

>
> **DID YOU KNOW?**
> A voluntary muscle working aerobically in a healthy individual generates about 280 mM ATP/min or about $1.69 \times 10^{23}$ molecules of ATP/min!

# TEST TIP

It is highly likely that you will need to understand the energy coupling of ATP, what phosphorylation means, and what phosphorylating a reactant accomplishes.

## Matter

**Matter**, which includes elements such as nitrogen, phosphorus, and sulfur, is necessary for the construction of all of the complex molecules in living systems, and it cycles through an ecosystem. Therefore, because living organisms are composed of matter, the basic rules of chemistry apply.

Matter is not lost, but instead is reincorporated into living tissues. For example, carbon—necessary for every class of organic molecules—enters a system as carbon dioxide that is then reduced to a sugar. This sugar is oxidized by living organisms. The carbon dioxide either is released or is incorporated into the tissues of organisms to be made available for reuse, either through decomposition after death or through the activity of photosynthesis.

All organisms require an input of energy from the environment, as well as the means to control the orderly use of that energy. Organisms generally convert the energy they obtain to ATP, the cell's "energy currency," which they use to power all life processes, including biosynthesis. Biochemical reactions, catalyzed by a large array of enzymes that are specific for each reaction, control biosynthesis—the chemical reactions that produce the macromolecules of which an organism's cells are composed.

Biological molecules include carbohydrates, lipids, proteins, and nucleic acids that have a variety of important functions. Water is the most abundant molecule in living organisms and possesses a variety of instrumental properties that result from its hydrogen bonding. Enzymes are proteins that act as catalysts to speed up biochemical reactions.

## Biological Chemistry

**Elements** are substances that cannot be further broken down by chemical means. These biologically important elements are found in all organisms. There are also other vital elements present in smaller amounts:

- Carbon (C), hydrogen (H), and oxygen (O) are found in all macromolecules.
- Nitrogen (N) is found in significant amounts in proteins and nucleic acids.
- Sulfur (S) is an element commonly found in proteins.
- Phosphorous (P) is prominent in nucleic acids.
- Sodium (Na), potassium (K), magnesium (Mg), and iron (Fe) are examples of important elements found in lesser quantities in most organisms.

An **atom** is the smallest unit of an element. The nucleus of an atom contains positively charged protons and neutral neutrons. The atomic number of an element is its number of protons, while two atoms that differ in mass form isotopes. Isotopes behave and are treated identically within living systems, but their differences have been exploited to solve many biological questions. The atomic mass of an atom is its number of protons plus neutrons.

The **electrons** of an atom surround the nucleus, are negatively charged, and have virtually no mass. An electron can become excited and leave its atom, and electrons can be passed from molecule to molecule. Electrons in the outermost energy level of an atom are involved in chemical bonding.

Atoms join together by chemical bonds to form compounds and molecules. Biological systems are chemical systems. All matter is made of atoms and the different forms and kinds of matter are the result of interactions between atoms. Compounds or molecules are made from atoms and interact, using or releasing energy, to form larger molecules or functional structures. This is the start of a hierarchy of interactions that ultimately leads to the interactions of all living things on Earth. Atoms interact based on their electron arrangements to form chemical bonds.

Valence electrons are the electrons in the outermost energy level of an atom and are responsible for an element's bonding behavior. Electrons move between atoms to form chemical bonds.

Some important chemical bonds are:

- **Ionic bonds** are formed when one atom transfers an electron to another atom. Ionic bonds are charge:charge interactions. For example, if an atom loses an electron, then it becomes a positively charged ion; if the atom gains an electron, then it becomes a negatively charged ion. It is important to remember that the attraction of opposite charged ions constitutes an ionic bond. An example of an ionic bond is found in the compound NaCl.

Figure 8-5. Example of an Ionic Bond: NaCl

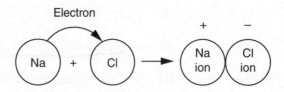

**Covalent bonds** are formed when two atoms share outer electrons. The majority of bonds within macromolecules are covalent. In living organisms, covalent bonds do not normally form or break without the aid of enzymes. If the electrons are shared equally between members of a bond, the bond is **nonpolar**. If the electrons are shared unequally, a **polar covalent bond** forms.

The bonds in a water molecule are examples of polar covalent bonds. Because of the unequal distribution of electrons, the hydrogen atoms carry a partial positive charge, and the oxygen atom carries a partial negative charge. This allows water molecules to form hydrogen bonds. The water molecule will be discussed further in the next section of this chapter.

Figure 8-6. Polar Covalent Bonds of a Water Molecule

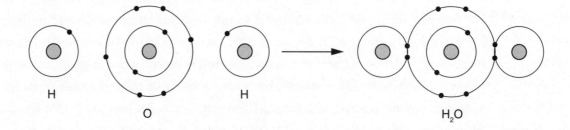

REDOX reactions are the most important type of chemical reactions within a biological system. REDOX reactions are also referred to as **reduction-oxidation reactions**. These are reactions that involve a partial or complete transfer of one or more electrons from one reactant to another. The element that "gave up" the electron(s) is **oxidized**; the one that "gained" the electron(s) is **reduced**. Often the electron transfer is really a shift from a less **electronegative** partner to a more electronegative partner, or vice versa. This shift is what happens in energy-rich carbohydrates. Hydrogen atoms and their electrons are removed from the carbohydrate and bonded to oxygen. The hydrogen atoms and their electrons have shifted from a less electronegative carbon partner to a more electronegative oxygen partner. This shift releases energy and is the basis for biological energy transfers. The opposite shift is what happens in photosynthesis, when electrons shift from a more electronegative partner (oxygen in water) to a less electronegative partner (carbon in carbon dioxide). This shift does, however, require an input of energy.

# TEST TIP

The most important biological reactions are REDOX reactions. They often involve the shift of electrons with their hydrogens, as from glucose to oxygen in cellular respiration. So, if asked about what molecules are oxidized in a reaction, the reactant that loses the hydrogen is the one that is oxidized and the one that gains the hydrogen (with its electron) is reduced.

## Water

Organisms are mostly water, and the reactions of biological molecules take place within water. Water has many significant characteristics important to living organisms, primarily because of its **polarity and hydrogen bonding**. In a water molecule, the very electronegative oxygen exerts a stronger hold on the shared electrons of hydrogen within a water molecule. The result is a polar molecule, a molecule with partial charges. If water molecules are brought together, the poles interact through hydrogen bonding, making water molecules cohesive.

**Hydrogen bonding** is a weak interaction between the hydrogen of one molecule with a more electronegative member of a neighboring molecule. Hydrogen bonding functions to stabilize many biological molecules, such as DNA and proteins.

Figure 8-7. Hydrogen Bonds Form Between Water Molecules

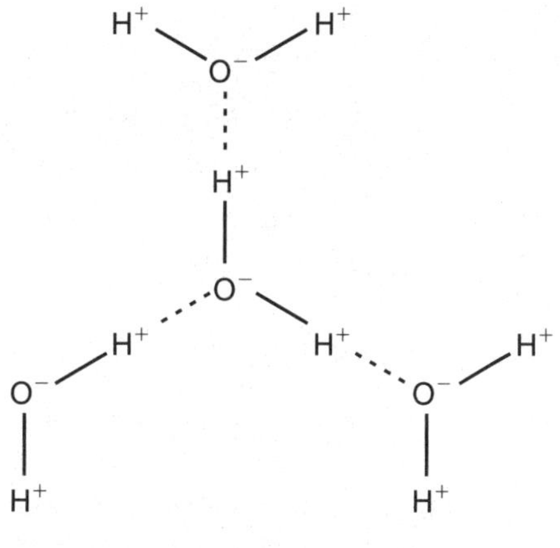

Dotted lines represent hydrogen bonds between water molecules.

Water is an *aqueous solvent* in which many biochemical reactions can take place—within a cell and in its immediate environment (the space between cells in a multicellular organism).

The pH of the aqueous environment inside a cell and its organelles influences many biological activities such as the shape of proteins, the creation of proton gradients across membranes, and the speed at which enzymes catalyze reactions. Biological systems must maintain acidity within narrow limits and have evolved mechanisms to do this.

> **DID YOU KNOW?**
>
> Water is considered so essential to life that, in any exploration of extraterrestrial planets for evidence of life, water is the first molecule that scientists look for to confirm the existence of life, past or present.

- **Acidic solution**—contains more $H^+$ than $OH^-$
- **Basic solution**—contains more $OH^-$ than $H^+$

The ability to interact with water has consequences for the ultimate structure of many biological systems. Substances can be classified as **hydrophilic** (water loving) or **hydrophobic** (water fearing), that is, a substance can interact with water or not. For example, the precise conformation of proteins is often due to the presence of hydrophobic or hydrophilic amino acids interacting with the watery environment of the cell.

Because water has a high specific heat capacity, it cools down and heats up slowly, allowing for temperature stability of organisms and the aqueous environments in which many organisms live. Water has many important properties that make it a necessary molecule for many organisms. See Table 8-2 for more specifics.

Table 8-2. Properties of Water

| Property | Description | Example |
| --- | --- | --- |
| Adhesion | H-bonding between water and dissimilar molecules | Imbibition contributing to seed germination |
| Cohesion | H-bonding between water molecules causes them to "stick together," giving water substantial surface tension. | Water striders are able to walk on water. |
| High specific heat | The amount of heat energy necessary to raise the temperature of water is high. | Large bodies of water absorb and release heat slowly and can moderate localized temperatures. |
| High heat of vaporization | The amount of heat energy necessary to move from the liquid to the gaseous state is high. | Animals can cool themselves by sweating. |

*(Continued)*

(*Continued*)

| Property | Description | Example |
|---|---|---|
| Expansion upon freezing | As water molecules cool and slow down, they become locked into a structure that is less dense than liquid water. | Ice floats, so it acts as an insulator on bodies of water. |
| Excellent solvent | Water dissolves most substances. | Biological reactions take place in water. |

# Monomers and Polymers

**Monomers** are the building blocks of larger macromolecules called *polymers*. **Macromolecules** are large molecules that fall into four categories: carbohydrates, lipids, proteins, and nucleic acids.

**Condensation** reactions are responsible for the biosynthesis of polymers from monomers with the *removal of water*. See Figure 8-8 below for an example.

Figure 8-8.  Condensation Synthesis of a Polymer

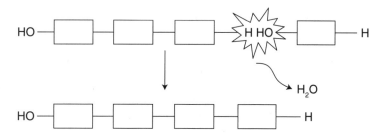

**Hydrolysis reactions**, conversely, break down polymers to their monomers with the addition of water, and in Figure 8-9, we see it is the reverse of the reaction shown in Figure 8-8.

Figure 8-9.  Hydrolysis of a Polymer

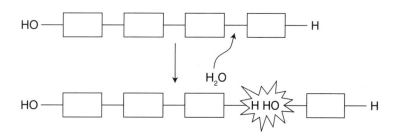

# Biological Molecules

Interactions occur at every level within the biological world. Atoms interact to become molecules; molecules interact to form ever-larger molecules called macromolecules. A hierarchy of organization then progresses to organelles to cells to tissues to organs to systems in a complex, multicellular organism. At each new level of organization, properties emerge that were not present at the previous level of organization.

Organic compounds are carbon-based compounds. Carbon has the ability to make four bonds in all directions making it suitable for forming the backbone of large, complex, branching molecules termed macromolecules.

The shapes of organic molecules are very important to their function. For example, biological molecules often exist in several forms or shapes called isomers. Each shape will behave chemically differently. For example, glucose isomerization to fructose in glycolysis makes the molecule ready to split into two three-carbon molecules, a key step in glycolysis. Often the shape or behavior of a molecule is affected by a functional group. A functional group changes the property of a molecule or of an area of a macromolecule.

There are four classes of biological molecules: carbohydrates, lipids, proteins, and nucleic acids. All of these molecules carry out a variety of functions to help the organism function.

## Carbohydrates

Carbohydrates are molecules that have the general formula $CH_2O$. They are made up of monomer units called monosaccharides that function to provide energy and structure. The most important monosaccharide is glucose ($C_6H_{12}O_6$). In fact, sugars are key metabolites used in the synthesis of other organic molecules, as well as the substrates of glycolysis and the products of photosynthesis. The specific orientation of a carbohydrate and its secondary structure are determined, in fact, by their particular bonding between carbohydrate subunits. Therefore, two forms of glucose are called **stereoisomers** because they contain the same atoms, but differ in orientation. See Figure 8-10 for an example. The two different orientations make a difference in how they can and will be used by an organism.

Figure 8-10.  Stereoisomers of Glucose

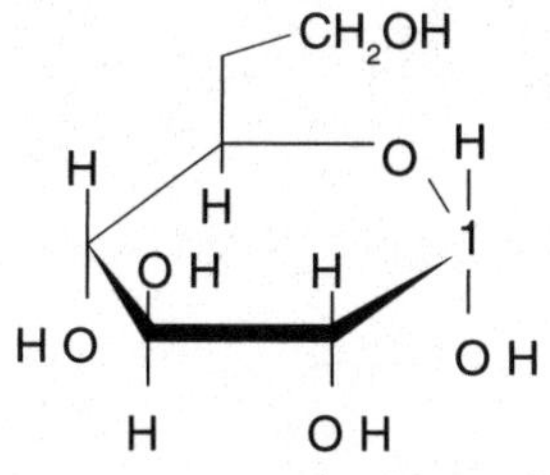

Alpha (α) Glucose 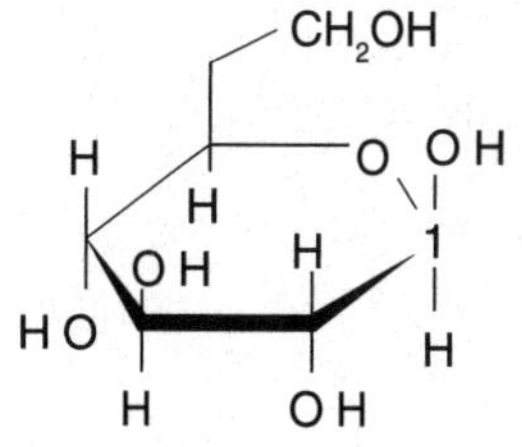                Beta (β) Glucose

## Lipids

Lipids are also composed of C, H, and O, but with a higher ratio of H:O than carbohydrates. In fact, in the form of fats and oils, lipids provide twice the energy as carbohydrates. Lipids exhibit several different kinds of structures, but their common trait is their insolubility in water.

Fats and oils are formed from a glycerol molecule bonded to one to three fatty acid chains. In general, fats are produced by animals and are saturated. Oils are produced by plants and are unsaturated. The difference between saturated and unsaturated fats is the presence or absence of double bonds between one or more of the carbons of the fatty acid chains. Saturated fats are solid at room temperature and unsaturated fats are liquid at room temperature due to the presence of unsaturated fatty acids.

Figure 8-11.  Saturated Fatty Acid

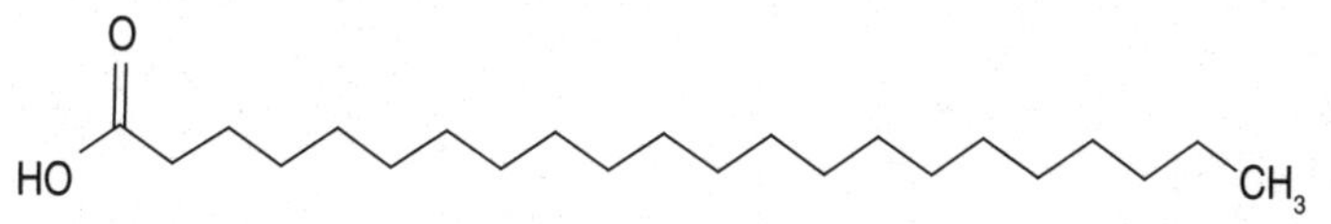

**Unsaturated fatty acids** have one or more double bonds and are more likely to be fluid at room temperature.

Figure 8-12.  Unsaturated Fatty Acid

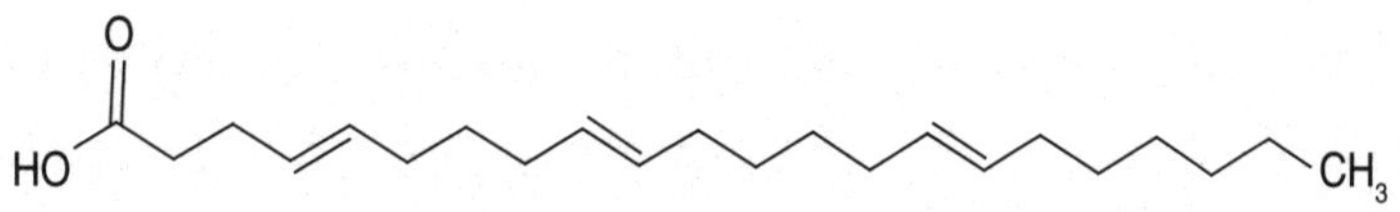

**Phospholipids** are an especially important group of lipids. These compounds have one nonpolar hydrocarbon tail in a triglyceride replaced by a polar phosphate group. In an aqueous environment, these kinds of lipids naturally form a bilayer and make up the cell membrane.

Figure 8-13.   Phospholipid

**Wax and steroids** (including cholesterol) are lipids with more complex structures that have a variety of functions.

Figure 8-14.   Cholesterol

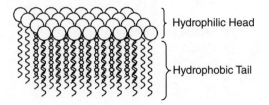

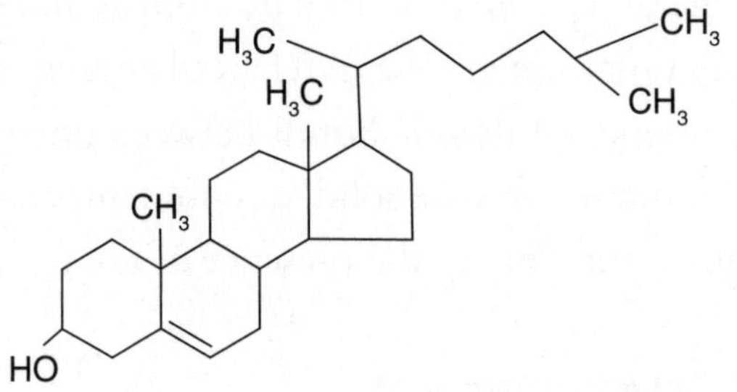

# Proteins

Proteins are polymers of amino acids that carry out most of the metabolic functions of an organism. Proteins are the workhorses of living organisms performing or mediating most of the metabolic functions of a cell. A protein's effectiveness is based on its precise conformation. The overall shape of a protein is determined by the unique sequence of amino acids as directed by instructions carried on by DNA, as well as the environment in which it is made. There are 20 amino acids, all of which have a carboxylic acid group, an amine group, and a side group that can be polar, nonpolar, or electrical in nature. A protein's shape has up to four levels of organization:

- The **primary structure** refers to the specific sequence of amino acids in a polypeptide;

- The **secondary structure** refers to the initial folding patterns of certain lengths of the polypeptide chain, such as alpha helices and beta sheets;

- The **tertiary structure** refers to the overall shape in which a polypeptide eventually folds; and

- The **quaternary structure** arises from the association of two or more folded polypeptides to form a multi-subunit protein.

A protein's form is dependent primarily on the order of amino acids as directed by DNA through RNA and on the environment it is in. If a protein loses its form because of environmental factors, it is said to be **denatured**. Denaturing may be permanent, as through the action of high heat, or temporary, based on the environment. Proteins may be formed in an inactive form that becomes active with a change in the environment. This change in protein conformation due to changes in the environment is the major way that proteins carry out their tasks.

**Amino acids** share the same basic structure as proteins, but the R group is different:

Figure 8-15.  Example of an Amino Acid

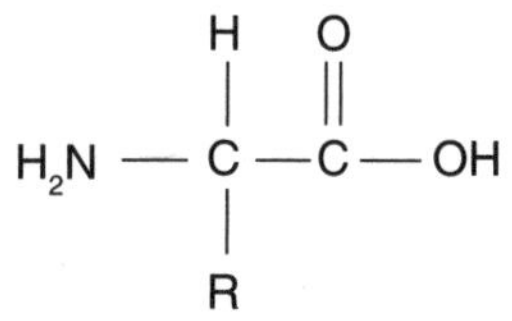

$$H_2N - \underset{\underset{R}{|}}{\overset{\overset{H}{|}}{C}} - \overset{\overset{O}{\|}}{C} - OH$$

Amino acids are connected by a linear sequence through the formation of peptide bonds by dehydration synthesis. They contain a central carbon atom covalently bonded to four atoms or functional groups arranged as follows:

- One of the four is always a hydrogen atom

- A carboxyl functional group (acidic) –COOH and an amine functional group (basic) –NH2

- The fourth component is a variable R group, which is different for each amino acid

# Nucleic acids

Nucleic acids are polymers of nucleotides. DNA is the double-stranded macromolecule that carries the instructions necessary for maintaining all of life's activities. The DNA nucleotide bases are adenine, thymine, cytosine, and guanine. RNA is single-stranded and processes the information from DNA. Its bases are adenine, uracil, cytosine, and guanine. Nucleotides are also components of energy transfer molecules. ATP is the most important energy transfer molecule in biological systems. It is a nucleotide with three phosphate groups attached. The third phosphate group is readily transferable to components of a reaction, energizing the reactants enough to enter a reaction. The coenzymes $NAD^+$ and $NADP^+$ are coenzymes that are built from nucleotides that act as electron acceptors in cellular respiration and photosynthesis, respectively.

Figure 8-16. Nucleotide

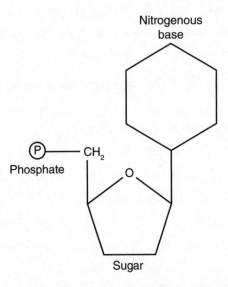

Table 8-3. Types, Examples, and Functions of Biological Molecules

| Type of Biological Molecule | Examples | General Functions |
| --- | --- | --- |
| Carbohydrates | Monosaccharides (sugars) Glucose | Energy; building blocks of other carbohydrates |
| | Deoxyribose and ribose | DNA and RNA |
| | Polysaccharides Starch and glycogen | Energy storage |
| | Cellulose | Plant cell wall structure |

*(Continued)*

(*Continued*)

| Type of Biological Molecule | Examples | General Functions |
|---|---|---|
| Lipids | Fats<br>Phospholipids<br>Waxes<br>Steroids (cholesterol) | Energy storage<br>Plasma membrane structure<br>Physical protection<br>Hormones (part of cell membranes) |
| Proteins | Enzymes<br>Other proteins | Biochemical catalysts<br>Structure, movement, signal reception, etc. |
| Nucleic Acids | DNA<br>RNA<br>ATP | Storage of genetic information<br>Converts genetic information into proteins<br>Energy currency of the cell |

Keep in mind that variations within these biological molecules allow for cells and organisms to possess a much wider variety of functions, such as having different types of hemoglobin or different phospholipids on a cell membrane.

## TEST TIP

Biological molecules are great examples of the "structure and function" theme. Be sure to use these examples in essay questions if appropriate.

# Enzymes

**Enzymes** are proteins that act as catalysts to speed up biochemical reactions. The function of enzymes is to lower the *activation energy* of a reaction. The activation energy of a reaction is the energy required to initiate a chemical reaction.

Figure 8-17. Energy Profile for an Enzyme

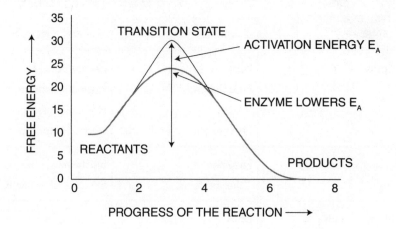

The enzyme combines with the substrate or molecule on which that the enzyme will work. The shape of the enzyme's reactive site matches the shape of the substrate molecule.

Figure 8-18. Enzyme and Substrate Molecules

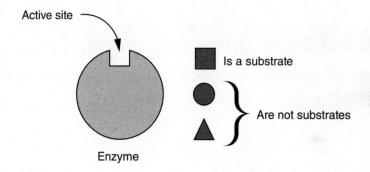

When the enzyme and substrate are joined, a catalytic reaction takes place, forming a product. The enzyme can be recycled and used for later reactions.

Enzymes are affected by pH, temperature, and substrate concentration. Enzymes have a pH and temperature optima at which enzyme activity is greatest. Also, as a substrate concentration increases, the speed at which the reaction occurs increases up to a maximum level at which all enzyme molecules are processing substrate molecules as quickly as possible.

**Cofactors** and **coenzymes** can also affect enzyme function. Sometimes the interaction between them changes a structural change and, therefore, the enzyme's activity rate. Enzymes may also only become active when all necessary coenzymes and cofactors are present.

# Chapter 9

# Photosynthesis and Cellular Respiration

## Photosynthesis

**Photosynthesis** occurs in all photosynthetic autotrophs including plants, algae, and photosynthetic prokaryotes. In eukaryotes, photosynthesis occurs in chloroplasts. In prokaryotes, it occurs in the plasma membrane and in the cytoplasm.

Photosynthesis is a REDOX reaction that oxidizes $H_2O$ and reduces $CO_2$. The overall equation of photosynthesis is:

$$6CO_2 + 6H_2O + \text{light energy} \rightarrow C_6H_{12}O_6 + 6O_2$$

Photosynthesis is affected by a variety of environmental factors. In eukaryotes, each phase of photosynthesis takes place in the **chloroplasts**, which is found in the mesophyll of plant leaves. A chloroplast is a double lipid bilayer that contains a clear fluid called **stroma** and stacks of green thylakoid membranes called **grana**. The $CO_2$ that is a necessary raw material of photosynthesis enters the leaf through pores called **stomata**, which are usually found on the underside of the leaf.

Figure 9-1. Chloroplast

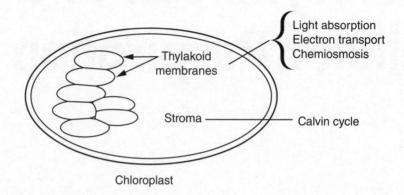

The $O_2$ waste generated by photosynthesis is released and exits the leaf through the stomata as well. The opening and closing of the stomata is accomplished by the actions of two guard cells, one on either side of the stomate. The stomata open and close based on turgidity. If the guard cells are turgid, the cells will be swollen with water and the stomate will open allowing gases to pass both ways. If the guard cells are flaccid, they will flatten and close the stomate.

> **DIDYOUKNOW?**
>
> A single mature tree can absorb carbon dioxide at a rate of 48 lbs/year and release enough oxygen back into the atmosphere to support 2 human beings. A 100-ft tree with an 18-inch-diameter base can produce 6,000 pounds of oxygen per year.

## The Two Steps of Photosynthesis

Photosynthesis contains two steps. The first step is a light-dependent reaction in which light energy is absorbed and converted to the chemical energy of ATP and the reducing power of NADPH. The second step, the light-independent reaction, uses the ATP and NADPH to convert $CO_2$ to sugars using the Calvin cycle.

### Step 1: The Light-Dependent Reaction

The **light-dependent reaction** occurs in the *thylakoid* membranes of chloroplasts in eukaryotes and oxidizes water. The thylakoid color is due to the presence of chlorophyll, the predominant pigment found in plants.

The pigment molecules collect light energy. Chlorohyll a is the pigment responsible for capturing the light energy for photosynthesis. However, chloroplasts have other pigments that broaden the **action spectrum**, the spectrum of light that can contribute to photosynthetic activity.

**Chlorophyll b and carotenoids** are accessory pigments that absorb higher wavelength light energy and pass it on to chlorophyll a. The following graph shows the absorption spectra of photosynthetic pigments:

Figure 9-2. Absorption Spectra of Photosynthetic Pigments

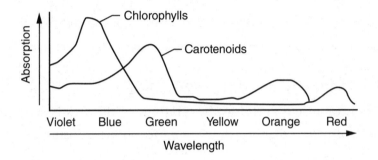

The light dependent reactions occur when **photons** of light excite the electrons of chlorophyll a found in the reaction center of **photosystems**. Photosystems are protein complexes with pigment molecules embedded. The pigment molecules absorb light energy of varying wavelengths, which excites their electrons. The energy of the excited electrons is passed along until it reaches the level of energy necessary to excite the electrons of chlorophyll a in the reaction centers of the photosystems. The electrons from the reaction center are excited and captured by an electron acceptor. The electrons do not return to their original position. The most important electron acceptor in photosynthesis is **NADP⁺**, a coenzyme.

Two photosystems work together to accomplish the light reactions of photosynthesis.

**Photosystems I and II** absorb energy at slightly different wavelengths. The electrons from photosystem I are excited and captured by an electron acceptor which will pass these electrons to NADP⁺ to form NADPH⁺ H⁺. The ejected electrons leave an electronegative hole in the reaction center of photosystem I.

The electrons from photosystem II are simultaneously excited and captured. Photosystem II electrons are passed down the electron transport chain (ETC) and fill the hole in photosystem I. The *very* electronegative hole in photosystem II is filled by electrons that were generated when an enzyme splits water into two electrons, two hydrogen ions, and an oxygen atom.

The lone oxygen atom is unstable and very quickly bonds with another oxygen atom when the enzymatic reaction of splitting a water molecule reoccurs, thereby generating $O_2$. As a result, the hydrogen ions become part of the reservoir of hydrogen ions in the stroma, and the electrons are fed into the electronegative holes in photosystem II, one at a time.

The **electron transport chain (ETC)** is an electrochemical gradient of hydrogen ions (protons) across the thylakoid membranes that undergoes redox reactions in a series. The electronegative compounds on the ETC are alternately reduced by their neighbor upstream on the chain, then oxidized by their downstream neighbor. At various points along the chain, $H^+$ ions are translocated from the stroma to the thylakoid space, creating an electrochemical gradient. ATP synthase is embedded in the thylakoid membranes and the $H^+$ move through the ATP synthase. The downhill flow of the hydrogen ions through ATP synthase phosphorylates ADP to ATP. This is called **chemiosmotic photophosphorylation**, or chemiosmosis, a process very similar to the ATP formation that occurs in the mitochondria during cellular respiration.

Electrons can take either a non-cyclical or a cyclical route. The primary difference between the two is that the cyclical flow of electrons produces more ATP and takes place because the Calvin-Benson cycle uses more ATP per mole than NADPH per mole, and hence replenishes the used ATP. In chemiosmosis, the movement of $H^+$ ions down their concentration gradient happen from inside the thylakoids to the stroma. As they do this, they pass through the enzyme ATP synthase, which causes the catalysis of ATP from ADP and $P_i$.

Figure 9-3.  ETC and Chemiosmosis of Photosynthesis

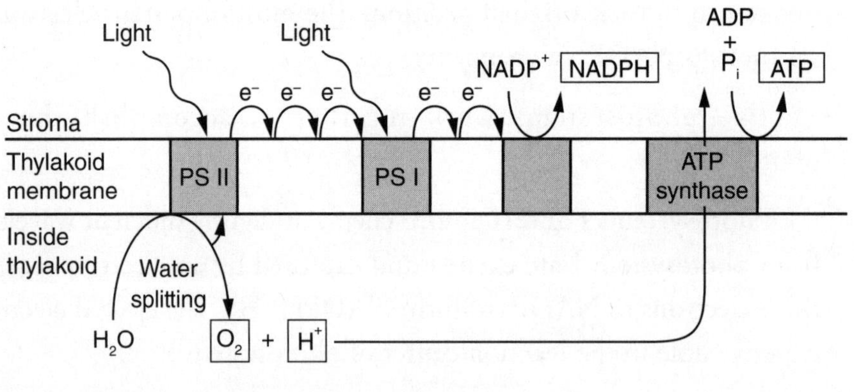

The important raw materials of the light reactions of photosynthesis are light energy and water. The important products of the light reactions of photosynthesis are

ATP and electrons to be used in the light-independent reactions. An important waste product of the light reactions is oxygen gas. Note that oxidizing water is very energy intensive. It was originally thought that carbon dioxide was the molecule that was split in photosynthesis.

## Step 2: The Light-Independent Reaction (The Calvin-Benson Cycle)

The second stage of photosynthesis takes place in the stroma and is the light-independent stage, which is often called the Calvin-Benson cycle after its discoverers. The products of the light reaction provide the energy, the hydrogen ions, and the electrons needed to reduce $CO_2$ to a 3-carbon molecule called glyceraldehyde-3-phosphate (G3P). To produce this three-carbon sugar, the *cycle will have to be repeated three times*. The light independent reactions take place in three stages:

1. **Carbon fixation**. The carbon dioxide is attached to a 5-carbon molecule (ribulose bisphosphate) and this splits into two 3-carbon molecules.

2. **Reduction**. Each of the 3-carbon molecules is phosphorylated using ATP generated in the light-dependent reactions, then reduced using the electrons captured in NADPH.

3. **Regeneration** occurs when the initial 5-carbon molecule uses ATP from the light-dependent reactions to restart the cycle.

A few things to note about the Calvin-Benson cycle:

- The cycle must occur three times for every G3P that is generated. It is very endergonic, and it is a cycle that begins and ends with the same molecules. Also, the end product is not glucose as you usually see written in the equation for photosynthesis.

- The Calvin-Benson cycle is the genesis of the chemical energy that becomes available to non-photosynthetic organisms.

- The reactions do not require light, and they take place in the stroma.

- Six turns of the cycle fix six carbons, representing one molecule of glucose, and six turns of the cycle require 18 ATP and 12 NADPH.

Figure 9-4.  The Calvin-Benson Cycle

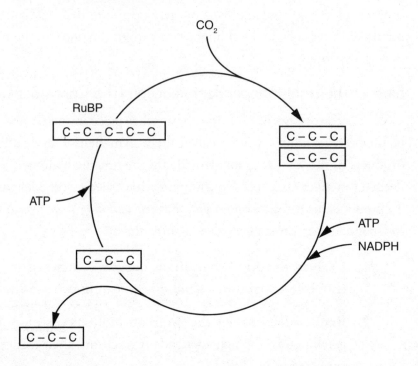

- Finally, oxygen gas is a competitor with carbon dioxide for the active site of ribulose bisphosphate carboxylase (RuBP), the enzyme that fixes $CO_2$ during the light-independent reactions that take place in the stroma. If the concentration of oxygen in the mesophyll of the leaf goes up, as it does on hot dry days, the Calvin-Benson cycle loses carbon. The $O_2$ displaces $CO_2$ in RuBP. When the cycle is supposed to release G3P, it releases a 2-carbon compound, which means the plant is losing carbon instead of gaining it. In addition, ATP energy was used but no product was gained. This process is called **photorespiration**. Some plants have adaptations that counter this by securing $CO_2$ when it is plentiful in order to keep the $CO_2$ to $O_2$ ratio high. $C_4$ and CAM plants are examples of such plants.

# TEST TIP

Know that molecules are oxidized and reduced in photosynthesis and cellular respiration and that these molecules are similar. Also, be able to compare the ATP production in the electron transfer chains of photosynthesis and cellular respiration.

# Fermentation and Cellular Respiration

**Cellular respiration** is the breakdown of glucose to produce energy (ATP) and organic intermediates used in the synthesis of the other organic molecules (amino acids, lipids, etc.) needed by the cell. Some form of cellular respiration takes place in nearly all organisms.

- *Glycolysis* is the oldest metabolic pathway, is virtually universal, and takes place in the cytoplasm of the cell.
- *Aerobic respiration*—the Krebs cycle, electron transport, and chemiosmosis—takes place in the mitochondria in eukaryotes.

The overall equation for cell respiration is:

$$C_6H_{12}O_6 + 6O_2 \rightarrow 6CO_2 + 6H_2O + \text{Energy (ATP)}$$

Although this equation is almost the reverse of the equation for photosynthesis, the two processes involve different enzymes and biochemical pathways, as well as different organelles.

Cells may utilize an anaerobic pathway (fermentation) that does not require $O_2$ or an aerobic pathway that does require $O_2$. Glycolysis is the first step of both pathways, as this step does not require $O_2$.

> **DID YOU KNOW?**
> Brain cells, skeletal muscle cells, heart muscle cells, and eye cells contain the most mitochondria per cell, as many as 10,000 mitochondria per cell. On the other hand, red blood cells have no mitochondria!

Aerobic respiration has three additional steps, the second of which requires $O_2$ as the final electron receptor of the electron transport chain.

- The Krebs cycle takes place in the matrix of the mitochondria.
- The electron transport chain takes place in the inner membrane of the mitochondria.
- Chemiosmosis takes place across the inner membrane of mitochondria.
- The following figure (Figure 9-5) shows the location of fermentation and the steps of cellular respiration in the mitochondrion:

Figure 9-5. Location of Fermentation and the Steps of Cellular Respiration

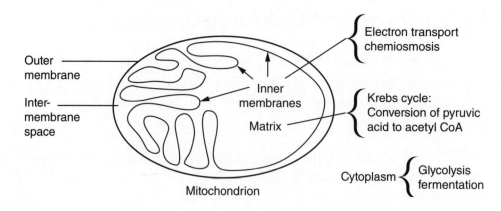

Anaerobic fermentation has one additional step following glycolysis that regenerates the oxidizing agent, $NAD^+$, to allow glycolysis to operate in the absence of $O_2$.

Both fermentation and cellular respiration are catabolic and involve oxidation-reduction reactions: The loss of electrons is oxidation (glucose to carbon dioxide); gain of electrons is reduction (oxygen to water); and electrons equal energy.

## The Four Parts of Cellular Respiration

**1. Glycolysis** is a ten-step metabolic pathway, catalyzed by a series of enzymes that split a glucose molecule into two molecules of pyruvic acid (pyruvate). This process is highly conserved and occurs in virtually every living organism. Glycolysis takes place in the cytoplasm and is divided into two phases: the energy-investment phase and the energy-yielding phase.

Essentially, glucose moves into a cell and is phosphorylated, rearranged, and phosphorylated again. This traps and destabilizes glucose and prepares it for splitting. ATP acts as an important allosteric control mechanism during this phase, so if ATP is plentiful, glycolysis will slow down. At the end of the energy investment phase, $NAD^+$ is reduced to NADH, one of the two major electron acceptors in cellular respiration; the other one is FAD, but to a lesser extent. Free energy is then released in the form of ATP—which comes from ADP and inorganic phosphates. This stage ends with two molecules of pyruvic acid that are transported to the cytoplasm for future oxidation.

### Figure 9-6. The Process of Glycolysis and its End Products

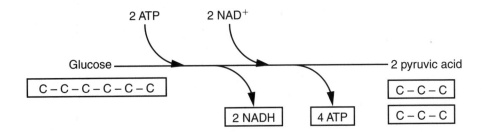

The important products of glycolysis are 2 net ATP, 2 NADH, and 2 molecules of pyruvic acid.

*Note:* At this point, if $O_2$ is absent, a series of reactions called fermentation will occur, but if $O_2$ is present, aerobic respiration will follow glycolysis.

*[handwritten: transition reaction]*

2. The **Shuttle Step** converts pyruvic acid to acetyl-CoA and occurs in the matrix of the mitochondria. It involves the following three important features:

- Coenzyme A (CoA) is added.
- Pyruvate is oxidized, producing NADH.
- The 3-carbon pyruvate is converted to the 2-carbon acetyl CoA, releasing a molecule of $CO_2$.

Each of the above molecules will enter the Krebs (Citric Acid) cycle.

### Figure 9-7. Conversion of Pyruvic Acid to Acetyl CoA

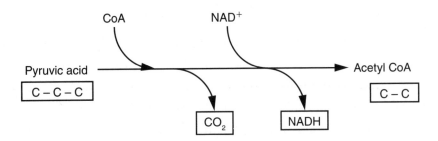

3. The **Krebs (Citric Acid) cycle** is a series of reactions that take place in the matrix of the mitochondria and continually regenerate one of its first reactants, oxaloacetic acid. The Krebs cycle produces the majority of NADH, $FADH_2$, and $CO_2$ (waste product) for cellular respiration. Three NADH and one $FADH_2$ molecules will be produced. Two molecules of $CO_2$ will be released, and one ATP will be formed. The key intermediate, oxaloacetate (OAA), is added to acetyl CoA to make citrate, which starts the entire Krebs cycle.

Figure 9-8. The Krebs Cycle

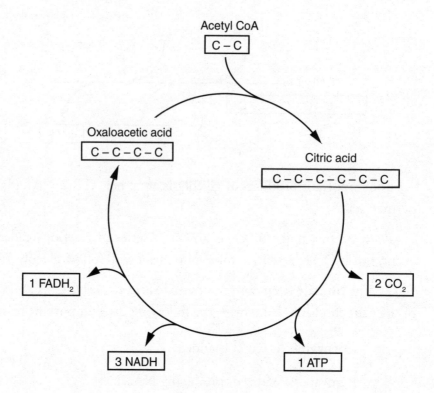

This part of cellular respiration will cycle twice for each molecule of glucose, so double the products will result.

4. **Electron Transport Chain.** The majority of the energy from aerobic respiration comes from oxidative phosphorylation, which is a result of the electron transport chain and the final stage of cellular respiration. This process takes place on the cristae of the mitochondria.

Many clusters of protein complexes are located in the cristae of the mitochondria and are arranged from least electronegative to most electronegative in an electron transport chain (ETC). The electrons oxidize from glucose, reduce $NAD^+$ to NADH and FAD to $FADH_2$. These pass the electrons to the first protein complex of the ETC. The neighboring protein complex oxidizes the first complex to become reduced, and this continues as the electrons are passed along the ETC.

Exerting the ultimate pull on the electrons is $O_2$ supplied by the respiratory and circulatory systems. At several points along the ETC, some of the electron transfers are coupled with the translocation of $H^+$ ions from the matrix to the intermembrane space. This makes the intermembrane space more positive and more acidic. The $H^+$ ions in the intermembrane space represent a proton motive force that cannot easily move through the inner lipid bilayer back into the matrix. Instead, embedded in the cristae are many

ATP synthase molecules that allow the H⁺ ions to flow down their concentration gradient. As this occurs, ATP synthase phosphorylates ADP molecules to ATP.

*Note:* In the ETC, no direct ATP is made; it must be coupled to oxidative phosphorylation via chemiosmosis (or the diffusion of H+ ions across the membrane—see above).

Cellular respiration produces a total of 38 ATP.

## TEST TIP

There is a good chance that at least one test question will require you to compare photosynthesis and cellular respiration. Be sure to keep in mind that plants carry out *both* photosynthesis and cellular respiration. Review the photosynthesis and cellular respiration chart (Table 9-1) below.

Table 9-1. Comparison Chart of Cellular Respiration and Photosynthesis

| Process | Takes Place in Cellular Respiration | Takes Place in Photosynthesis |
|---|---|---|
| Breakdown of glucose | Yes | No |
| Synthesis of glucose | No | Yes—Calvin Cycle |
| $O_2$ is released | No | Yes—light dependent reaction |
| $O_2$ is consumed | Yes—ETC and oxidative phosphorylation | No |
| Chemiosmosis | Yes—ETC | Yes—ETC |
| $CO_2$ is released | Yes—Shuttle Step and Krebs Cycle | No |
| $CO_2$ is consumed/fixed | No | Yes—Calvin Cycle |
| ATP is produced | Yes—glycolysis, Krebs Cycle, ETC, and Oxidative Phosphorylation | Yes—light dependent reaction |
| ATP is consumed | Yes—glycolysis initial investment | Yes—Calvin Cycle |
| Pyruvate as intermediate | Yes—glycolysis | No |
| NADH produced | Yes—glycolysis, shuttle step, Krebs cycle | No |
| NADPH produced | No | Yes—light dependent reaction |

# TEST TIP

You do not have to know any of the enzymes of each of the stages of cellular respiration or the number of ATP produced in each of the stages. But you should know that most of the ATP generated by cellular respiration is generated by oxidative phosphorylation and that the ultimate driver of the process is the presence of oxygen in the mitochondria. You should also remember where in the cell and/or mitochondria each stage of cellular respiration occurs.

## Fermentation

**Fermentation** provides a small amount of ATP in the absence of $O_2$. Fermentation oxidizes the NADH that was formed during the energy-yielding phase to $NAD^+$. This allows $NAD^+$ to be available for reuse in glycolysis so this process can continue. Products of fermentation are alcohol and $CO_2$ in microorganisms and lactic acid in animals.

**Glycolysis** is the first step of both aerobic respiration and fermentation. Two of the end products of glycolysis, pyruvic acid and NADH, can be processed anaerobically in the cytoplasm of certain cells. The second step of fermentation does not produce ATP directly. Rather, it generates $NAD^+$, which is required to keep glycolysis running and producing ATP.

**Lactic acid fermentation** includes glycolysis plus an additional reaction that generates $NAD^+$ and lactic acid. Certain fungi, bacteria, and muscle cells have special enzymes that carry out lactic acid fermentation. In vigorously exercising muscle cells, lactic acid fermentation provides ATP when the circulatory system cannot keep up with the oxygen demands of the muscle cells.

**Alcohol fermentation** includes glycolysis plus additional reactions that produce $NAD^+$, ethanol, and $CO_2$. Single-celled organisms, such as yeast and some plant cells, have special enzymes to carry out alcohol fermentation. Yeast is used in bread making because $CO_2$ gas causes bread to rise; the ethanol is removed by subsequent baking. Yeast is also used in beer making because it produces ethanol; $CO_2$ in an enclosed container produces carbonation.

Figure 9-9. Lactic Acid Fermentation and Alcoholic Fermentation

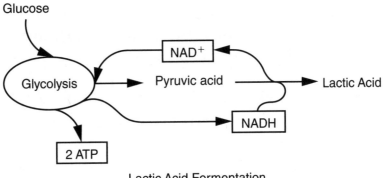

Lactic Acid Fermentation

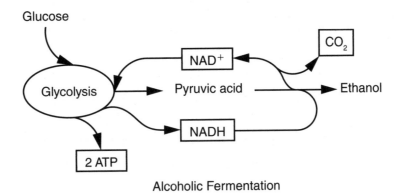

Alcoholic Fermentation

## Questions to Consider

1. What is the molecule that acts as a feedback mechanism for *cyclic* electron flow?

2. What is the function of both lactic acid fermentation and alcoholic fermentation?

3. Identify a key regulation point in the process of cellular respiration.

## Answers:

1. If there is an abundance of $NADP^+$, noncyclic electron flow will continue. If there is more NADPH, cyclic electron flow will occur.

2. Both of these processes take place in the absence of oxygen sometimes by organisms that obtain all of their ATP from glycolysis or sometimes by cells that are deprived of oxygen for one reason or another. The pyruvic acid that is the end-product of glycolysis is reduced by the NADH that was formed during glycolysis and this reduction consequently oxidizes NADH to $NAD^+$ so that it is recycled to continue accepting electrons. The function of fermentation (both lactic acid and alcohol) is to allow glycolysis to continue in the absence of oxygen.

3. A key regulation point takes place during the investment phase of glycolysis. ATP acts as an allosteric inhibitor of the enzyme that splits the 6-carbon fructose. If the cell has abundant ATP, ATP will inhibit the splitting process and slow glycolysis. This, in turn, slows the rest of cellular respiration. A cell will not expend energy, nor use valuable molecules, if there is no need to do so.

# Chapter 10

# Cellular Structure, Membranes, and Transport

## Prokaryotes and Eukaryotes

Cells are the basic structural and functional unit of living things. They are the fundamental unit of biology and the first stage at which life can appear. Cells are categorized into two main types: **prokaryotic** and **eukaryotic cells**.

**Prokaryotic cells** are simpler cells and have the following characteristics:

- No nucleus, only a nucleoid region with one, circular DNA

- No membrane-bound organelles

- A cell wall

- Contain a plasma membrane, cytoplasm, and ribosomes (location of protein synthesis)

- No histones or no formation of chromosomes

**Eukaryotic cells** include fungi, plants, protists, and animals; plant cells tend to be generally larger than animal cells. They have the following characteristics:

- Contain a nucleus, a nuclear envelope to protect DNA, nuclear pores to allow transport into and out of the nucleus, and linear DNA.

- Have membrane-bound organelles. Plants have chloroplasts, for example, where photosynthesis is carried out, and many plant cells have a large, central vacuole that is absent in most animal cells.

- Plants have rigid cell walls made of the polymer, cellulose, but animal cells do not have cell walls.

- Like prokaryotes, eukaryotes contain a plasma membrane, cytoplasm, and ribosomes (location of protein synthesis).

- Contain histones that form into chromosomes.

Figure 10-1. Animal Cell

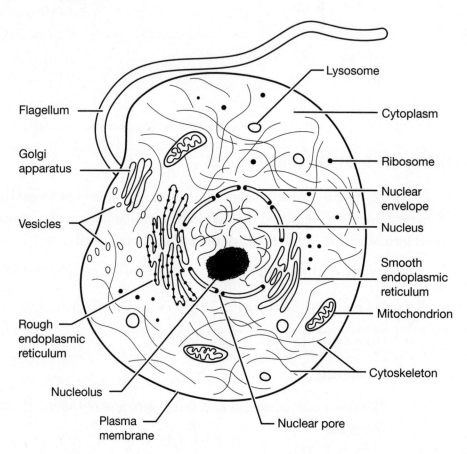

Figure 10-2. Plant Cell

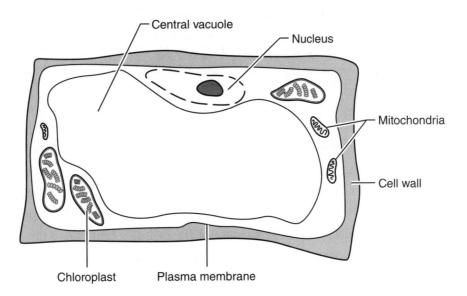

## The Endomembrane System and Eukaryotic Organelles

The **endomembrane system** of a eukaryotic cell is a series of cellular organelles that are connected by function and the kind of membrane that surrounds them. The endomembrane system compartmentalizes the inside of a cell so that there can be different conditions to allow for many different kinds of reactions to occur simultaneously. In addition, the endomembrane system increases the membrane surface area within a cell and thus increases the numbers and kinds of reactions that can occur within a cell. Some of the membranes are specialized for particular functions, and the interactions between membranous parts of the cell coordinate the activities of the cell.

Eukaryotic cells maintain **internal organelles** (including those in the endomembrane system) for specialized functions. Some of these organelles include the following:

- **Nucleus\***—surrounded by a double lipid bilayer called the nuclear envelope. This envelope is heavily populated by proteins and pores, each of which is ringed by protein complexes. The lipid bilayer, like all lipid bilayers, allows small, nonpolar molecules to pass but does not allow polar substances or large substances. The many pores, however, allow mRNA and ribosomal units to exit the nucleus to the cytoplasm. This membrane keeps the DNA separate

*Note: An asterisk (\*) indicates that the organelle is part of the endomembrane system.*

from the cytoplasm. Protein strands anchor the DNA to the nuclear envelope helping to keep it organized. The nucleus is also the site of chromosome storage (DNA) and RNA synthesis (transcription).

- **Nucleolus**—a dense, spherical area within the nucleus; the site of rRNA synthesis and ribosome production.

- **Mitochondrion** (plural, mitochondria)—a small organelle that is known as the powerhouse of the cell because it takes in nutrients, breaks them down, and creates energy for the cell. This process is known as cellular respiration. The mitochondrion is specifically shaped to be as efficient as possible. It is a small organelle because a cell may contain thousands of mitochondria, depending on the purpose of the cell. For example, brain cells and muscle cells contain more mitochondria than do many other cells in the body.

    A mitochondrion has two membranes. The outer membrane covers the organelle completely. The inner membrane is called the *cristae* and is folded to increase the surface area for electron transport, which directly requires oxygen. Similar surface area strategies for maximum space and efficiency are seen elsewhere, such as the microvilli in intestinal cells. The last part of the mitochondria is the fluid inside, which is called the *matrix*.

- **Endoplasmic reticulum (ER)\***—one continuous, highly folded and flattened rows of membranous sac with or without ribosomes attached. The inside of the sac is referred to as the lumen and separates substances from the cytoplasm. The endoplasmic reticulum can be either rough (contains ribosomes) or smooth (without ribosomes). The endoplasmic reticulum is the site of protein and membrane synthesis and also the detoxification of drugs.

- **Rough ER (RER)\***—contains ribosomes that translate proteins meant for secretion and attaches them directly to the RER and thread the translated polypeptide directly into the lumen of the RER, where modifications to the polypeptide chain can occur. These changes may be the addition of polysaccharide or other side chains. RER will be very abundant in cells that synthesize and secrete proteins. The RER also makes more of itself and generally maintains the cell's membranes.

> **DID YOU KNOW?**
> More than one-half of the cell's total membrane is endoplasmic reticulum.

---

*Note: An asterisk (\*) indicates that the organelle is part of the endomembrane system.*

- **Smooth ER (SER)\***—produces lipids from building blocks that were delivered to it. The SER produces vesicles and metabolizes toxic materials in the liver; in muscle cells, it sequesters calcium ions that enable muscle contractions. A protein in the lumen of the RER will move to the SER and bud off into a vesicle to be transported to the Golgi apparatus. Lipids produced within the SER will move in a vesicle to the Golgi apparatus.

- **Ribosomes\***—tiny organelles that do not have a membrane; instead, they contain rRNA and protein and are bound to the RER or float freely in the cytoplasm. They function as sites of protein synthesis.

- **Golgi apparatus\***—a series of flattened, membranous sacs with a receiving side and a shipping side. The receiving side of the Golgi apparatus merges with the vesicles shipped from the endoplasmic reticulum. The proteins that are moved through the Golgi stacks are progressively modified, mostly by addition and trimming of carbohydrate chains. Each sac of a Golgi apparatus is specialized to modify the molecules it receives. The last sac in a Golgi apparatus buds off a vesicle around the molecule to be shipped and targets its final location. When the vesicle meets the cell membrane, it merges with the membrane and what was inside the vesicle now faces the outside of the cell. The cell membrane has also been effectively enlarged by the addition of the vesicle membrane.

- **Chloroplast**—appears in plants and is the site of photosynthesis. Chloroplasts have double membranes plus thylakoids that are shaped like stacked coins in order to increase their surface area. Chloroplasts are where plastids can store starch or fat.

- **Vesicles**—small, spherical, and membranous sacs within a cell that are also abundant. They are surrounded by one membrane, and their function is to move substances from the ER to the Golgi apparatus and from the Golgi apparatus to the plasma membrane.

- **Lysosomes**—known as shipping vesicles; they are small, spherical, and surrounded by one membrane. They are also common in animal cells. Lysosomes hold enzymes capable of hydrolyzing all classes of organic molecules. When a cell has phagocytized a bacterium, for example, that vesicle will merge with a lysosome and the bacterium will be digested. Lysosomes also carry out the directed digestion that is part of apoptosis (programmed cell death). The lysosomes keep the enzymes contained or the cell in which it resides could be digested.

*Note: An asterisk (\*) indicates that the organelle is part of the endomembrane system.*

- **Vacuoles**—single membrane sacs that store water, food, wastes, and pigments. An important vacuole in plants is the large central vacuole. The central vacuole can take up as much as 90% of a plant cell's volume and is important to plant cell support. As mentioned earlier, contractile vacuoles enable single-celled protozoans to survive in a constantly hypotonic environment.

- **Peroxisomes**—sacs of enzymes, like the lysosomes, except the enzymes break down fatty acids to hydrogen peroxide. Then, this is further broken down into water and oxygen. In addition, peroxisomes degrade alcohol in liver and kidney cells.

> **TEST TIP**
>
> It is important to know the route of a secretory protein from its signal peptide sequence to its release from the cell via a vesicle, as well as the role of the endomembrane system in the preparation of a secretory protein.

# Cell Membrane Structure

All cells are characterized by a cell membrane, called the **plasma membrane**, that maintains the differences between the internal and external environments of the cell and allows substances to be transported in and out of the cell. The **cytosol** is a semifluid material between the cell membrane and the DNA within the cell. Cells are able to exchange materials with their surroundings, obtain and use energy, synthesize necessary molecules, grow and respond to signals, and reproduce.

Cells are small because of the metabolic constraints that go with exchange and reaction surfaces. A cell must maintain a high surface area to volume ratio to maximize the ability to take in raw materials and ship and get rid of waste products, as well as regulate processes vital to the cell. For example, a cube with a 1 cm side has a surface area of 6 cm$^2$ and a volume of 1 cm$^3$, a 6:1 surface area to volume ratio. If the side of the cube is increased to 2 cm, the ratio is 24 cm$^2$ to 8 cm$^3$, a 3:1 surface area to volume ratio. With the increased size of a three-dimensional figure, the volume increases disproportionately more than the surface area does.

This mathematics explains the often-seen phenomenon of *folding* in biological structures (e.g., mitochondria, Golgi apparatus, and endoplasmic reticulum). Folding increases surface area much more than it increases volume.

The plasma membrane is **selectively permeable**, meaning that it allows some substances to pass through it, but not others. This selectivity is a direct consequence of the membrane structure called the **Fluid Mosaic Model**. Essentially, the membrane is a mosaic of proteins that are embedded in or attached to the phospholipids.

The lipid portion of the membrane is composed mainly of phospholipids.

- *Structure*—The phospholipid portion of the membrane is composed of a polar **hydrophilic** (water loving) head and two **hydrophobic** (water fearing) fatty acid tails.

- *Environment*—The cytosol and the fluid outside of the cell (extracellular fluid) are both aqueous (watery) environments. So, the phosphate head interfaces with the watery environment that is outside and inside the cell.

- *Arrangement*—The fatty acid tails of separate phospholipids turn toward each other, away from the watery conditions surrounding the cell. This arrangement naturally forms a three-dimensional bilayer membrane.

Figure 10-3. Fluid Mosaic Model

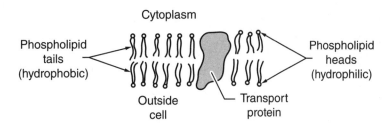

This membrane is not static. That is, the phospholipids and proteins are not attached to each other. They can and do move around, mostly laterally. The hydrophilic tails can be either saturated fatty acids or unsaturated fatty acids. Unsaturated fatty acid tails maintain a more fluid membrane packing than do saturated fatty acid tails.

**Cholesterol** is an important component of the cell membrane that helps maintain membrane fluidity when temperatures drop by preventing the close packing of the hydrophobic tails.

Another important aspect of the cell membrane is its hydrophobic core. This property enables small nonpolar molecules such as $O_2$ and $CO_2$ to cross freely, but it inhibits the crossing of ionic, polar, and large molecules.

**Embedded proteins** in the cell membranes can be hydrophilic, with charged and polar side groups, or hydrophobic, with nonpolar side groups. There are two types of embedded proteins, namely:

- **Integral proteins** are transmembrane proteins (i.e., they extend the entire way through the membrane) with hydrophilic and hydrophobic components. The reason why an integral protein that extends through the membrane must have hydrophobic components is so that it can pass through the hydrophobic core of the membrane.

- **Peripheral proteins** bind to integral proteins on the outside of the cell membrane. They reside close to the cell membrane and are often found in close association with a transmembrane protein.

The function of membrane proteins includes the following:

- transport
- enzymatic activity
- signal transduction and cell communication

Table 10-1. Functions of Membrane Proteins

| Protein Function | Description | Example |
|---|---|---|
| Channel proteins | Allows easy passage of a particular water soluble ion or molecule. | Aquaporins that allow easy passage of water |
| Carrier/transport proteins | Interacts with particular molecules. ATP energy may be necessary. | Sodium potassium pump |
| Cellular recognition | These proteins are usually glycoproteins. They have a carbohydrate chain attached. | MHC glycoproteins. These are different for each individual. |
| Enzymatic proteins | Catalyze specific reactions. | Adenylate cyclase is involved in ATP metabolism. If it is disrupted by cholera, severe diarrhea occurs. |
| Receptor proteins | Interacts with a specific signaling molecule. | Receive growth factor signals from neighboring cells and transduces the signal. |
| Intercellular joining/ communication | Allows cells to be locked together in sheets and some allow passage of material between cells. | Tight junctions and gap junctions. Plasmodesmata in plants. |
| Electron transfer proteins | Transfer electrons on an electron transfer chain. | The electron transfer chain embedded in the cristae of the mitochondria and thylakoid membrane of the chloroplast. |

**Cell walls** are nonliving structures that provide protection and support for a variety of different organisms. Among eukaryotes, all plant cells, as well as many protistans and fungi, have cell walls, though they are composed of different molecules. Many bacteria also have cell walls.

- Plant cell walls are mostly cellulose, an indigestible carbohydrate.

- Fungal cell walls are composed of chitin, a structural polysaccharide that contains nitrogen.

- In bacteria, the cell wall may be composed of peptidoglycans, peptide groups that crosslink polysaccharides.

### TEST TIP

You must know that the polar/nonpolar aspects of the cell membrane contribute to its selective permeability. Relate its structure to what kinds of molecules can easily cross the membrane versus what cannot so easily pass through the membrane.

## Cell Membrane Transport

The function of the cell membrane is to maintain homeostasis, and the major way of doing this is by controlling chemical movement into and out of the cell. Movement across the cell membrane takes two forms: passive transport and active transport.

### Passive Transport

**Passive transport** does not require the cell to use ATP energy and plays a role in both the import of resources and the export of waste. The two kinds of passive transport are diffusion and osmosis.

**Diffusion** occurs when a substance (gaseous or liquid molecules) moves down its concentration gradient from an area of higher concentration to an area of lower concentration. Substances that move are small, uncharged molecules (such as carbon dioxide and oxygen). If more than one substance is being diffused, each substance moves according to its own concentration gradient. Substances move directly across the lipid bilayer and spread out evenly.

Figure 10-4. Diffusion

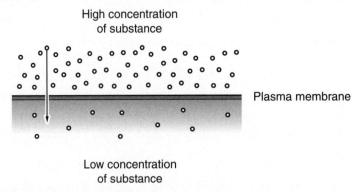

**Facilitated diffusion** moves charged molecules (e.g., potassium ions) and larger molecules (e.g., glucose) into and out of the cell. As with diffusion, facilitated diffusion moves a substance down the concentration gradient from an area of higher concentration to an area of lower concentration without the use of ATP. However, unlike diffusion, the substance moves with the help of carrier proteins or through a channel protein. Channel proteins allow substances to pass through the protein, while carrier proteins change their shape with the movement of their specific substance.

Water movement is often facilitated by specialized transport proteins called **aquaporins**. Examples of facilitated diffusion include glucose transport and $Na^+/K^+$ transport.

Figure 10-5. Facilitated Diffusion via Carrier Protein

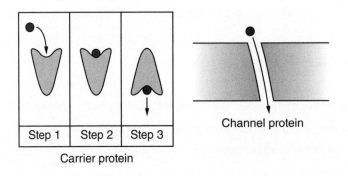

**Osmosis** is the diffusion of water and causes the water to move across the plasma membrane from a hypotonic solution to a hypertonic solution. Therefore, if water can move through a cell membrane but a solute cannot, three possibilities of solution concentration can occur: isotonic solution, hypotonic solution, or hypertonic solution.

An **isotonic solution** occurs when the cell and its surroundings have the same solute concentration inside the cell as it does outside the cell. In this situation, there will be *no net movement* of water in or out of the cell; this is called **dynamic equilibrium**. This represents a normal situation for cells without a wall; for a cell with a wall, the cell will be flaccid.

> **DIDYOUKNOW?**
> Pygmies are short not because they do not make enough growth hormone but because their cell membranes do not have the receptor protein for growth hormone.

A **hypotonic solution** is when the solute concentration inside the cell is greater than outside the cell. Water will move *into* the cell. If the cell is not bound by a cell wall, the cell will potentially **lyse**, or rupture. For a cell that is bound by a cell wall, the water will flow into the cell and the cell will swell up against the wall making the cell turgid. This is a positive effect for plant cells.

A **hypertonic solution** occurs when the solute concentration outside of the cell is greater than inside the cell; water will move *out* of the cell. Whether there is a wall or not, the cell will shrivel.

It is important to remember that for all of the conditions, that the concentrations are relevant. That is, a solution that is hypertonic to a cell in one situation may not be hypertonic to a different cell.

In osmosis, the solute molecule is not able to cross the *selectively permeable* plasma membrane.

Figure 10-6. Osmosis: Hypotonic, Hypertonic, and Isotonic Cells

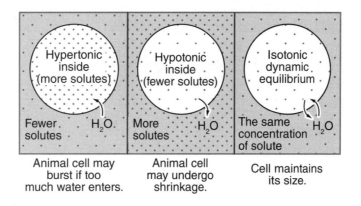

**Water potential** is the measure of the free energy of water, which is governed by pressure and solute concentration. Water moves from a higher water potential to a lower water potential. Solute concentration always lowers water potential because water molecules are tightly clustered around solutes, effectively reducing the number of free water molecules.

Water potential is calculated by the following equation:

Water potential ($\Psi$) = solute potential ($\Psi_s$) + pressure potential ($\Psi_p$)

Solute potential $\Psi_s$ (also the osmotic potential): The effect of solute concentration. Pure water has a solute potential of zero, but as solute is added, the value for the solute potential becomes negative.

Pressure potential $\Psi_p$: In a plant cell, pressure exerted by the cell wall to limit further water uptake; this pressure potential can be positive or negative.

The importance of water movement to and from a cell cannot be overstated. Many of the adaptations in the living world are responses to the potential for gaining or losing water to the environment.

*Example.* The presence of a contractile vacuole in the *Paramecium*. This unicellular organism is found in pond water, a generally hypotonic environment. The *Paramecium* has a contractile vacuole that pumps out the water that chronically moves into the cell. On the other hand, a marine fish maintains cells that are hypotonic to the seawater they live in. These animals can drink the seawater, but they possess specialized epithelial tissue that enables them to secrete excess solutes. One-celled marine organisms are generally isotonic to their environment.

Table 10-2. Summary of Cell Response to Relative Solution Concentration

|  | Hypotonic Solution | Isotonic Solution | Hypertonic Solution |
| --- | --- | --- | --- |
| Animal cell | Swell, lyse | Normal | Shrivel |
| Plant cell | Swell, turgid | Flaccid | Plasmolyze (shrivel) |

## Active Transport

**Active transport** requires the cell to use ATP and involves the movement against the concentration gradient. In other words, the solutes move from a lower concentration to a higher concentration; this is why active transport requires the use of ATP.

Specific membrane proteins are used in active transport as well. The process of active transport is critical to setting up membrane potentials and concentrating scarce raw materials.

Figure 10-7. Comparison of Active and Passive Transport Processes

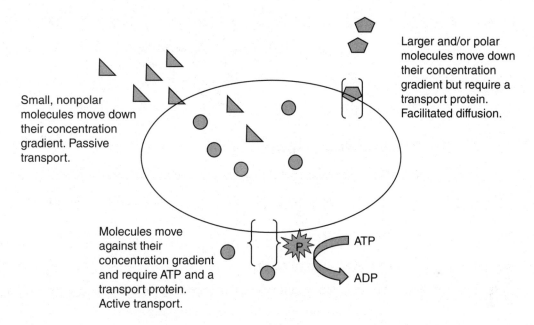

The **Na⁺/K⁺ pump** is an example of a protein that uses active transport to move ions through the cell's membrane. It is the action of the Na⁺/K⁺ pump that sets up the action potential in nerve cells.

Figure 10-8. The Na⁺/K⁺ Pump

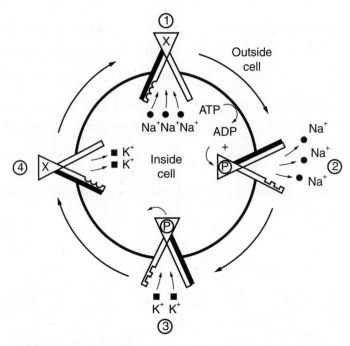

**Explanation of Figure 10-8.**

*Step 1:* Three Na⁺ ions in the cytoplasm move into the protein.

*Step 2:* Through phosphorylation, the protein changes shape, and the three Na⁺ ions are released outside the cell.

*Step 3:* The protein's changed shaped increases its affinity for K⁺. So, two K⁺ ions now move into the protein.

*Step 4:* Through dephosphorylation, the protein resumes its orginal shape, and the two K⁺ ions are released into the cytoplasm.

The **hydrogen pump** is another example of active transport that is used to establish a hydrogen ion gradient. In plants, proton pumps are set up in a manner similar to the Na⁺/K⁺ pump using ATP and a membrane protein to pump hydrogen ions from inside a cell to the outside of the cell. This gradient of hydrogen ions flows down through another, separate membrane protein that also has a site for co-transporting another molecule at the same time. This is an indirect active transport moving a molecule against its concentration gradient, using the concentration gradient of a separate solute.

**DIDYOUKNOW?**

One-third of the body's energy is expended on sodium-potassium pumps in the body. Each pump can move over 300 Na+ ions/sec.

Two active transport mechanisms that are also important to the maintenance of cellular homeostasis are **exocytosis** and **endocytosis**. They move large molecules of food particles across the plasma membrane with the expenditure of ATP. **Exocytosis** fuses the vesicles and molecules with the plasma membrane. It secretes the materials to the outside of the cell. In **endocytosis**, the cell takes in the molecules via vesicles that fuse with the plasma membrane. Two specific forms of endocytosis include **pinocytosis** that is used primarily for the uptake of extracellular fluids, and **phagocytosis** that is used for the uptake of solids. Phagocytosis is very important to the immune system for getting rid of invading pathogens.

Figure 10-9. Endocytosis and Exocytosis

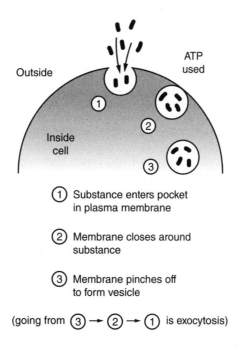

① Substance enters pocket in plasma membrane

② Membrane closes around substance

③ Membrane pinches off to form vesicle

(going from ③ → ② → ① is exocytosis)

# TEST TIP

It is very important to understand the use of electrochemical gradients to the maintenance of homeostasis. The sodium-potassium pump is the classic example of this and would be a good one to be very familiar with. You will not necessarily be asked about this pump only, but it provides you with a good example to talk about or draw from to apply to another mechanism.

## Questions to Consider

1. In a recent study, a large group of middle-aged males were divided into a control group and an experimental group. The experimental group maintained a low-cholesterol, high poly-unsaturated diet. Over the years of the study, the experimental group's blood cholesterol levels went down and the rate of arterial disease and heart disease were lower than those of the control group, not fed a low fat diet. However, even though the death rate due to heart attacks was much lower in the experimental group than in the control group, the overall death rate in the experimental group was significantly higher than that of the control group. The increased death rate was due to an increased cancer rate. Based on your knowledge of the cell, propose a hypothesis to explain this phenomenon.

2. A receptor protein destined for the cell membrane is shipped in a vesicle from the Golgi apparatus. What will be the orientation of the receptor in the vesicle so that the receptor is oriented correctly in the cell membrane?

3. Give examples of cells in the body that would have an abundance of each of the following endomembranes:

    (a) Rough endoplasmic reticulum
    (b) Smooth endoplasmic reticulum
    (c) Lysosomes

4. A red blood cell is dropped into a beaker of salt water and shrivels. What are two things that you infer from this description?

## Answers:

1. This study was actually done as part of the Framingham study on cardiovascular disease. One of the hypotheses for the increased death rate due to cancer was that the lack of cholesterol in the diet of the experimental group contributed to a lower quality cell membrane, making the cell more vulnerable to disease. It was also thought that the lower cholesterol contributed to a higher percentage of unsaturated fats, which form free radicals, highly reactive molecules that degrade other molecules.

2. The protein must be oriented so that the receptor part is outside the cell. This means that it must be oriented towards the inside of the vesicle. Then when the vesicle merges with the membrane, the receptor will be facing out.

3. (a) RER is found in cells that prepare proteins for secretion. The cells of the pancreas that prepare and release digestive enzymes and liver cells are examples of cells that would have a lot of RER.

   (b) SER would be very abundant in the muscles because of their role as a reservoir of calcium ions, and SER would be present in liver cells because of their role in detoxification of drugs and poisons.

   (c) Lysosomes are found in the macrophage cells of the immune system.

4. Because the cell shriveled, you can infer that the solution in the beaker is hypertonic to the red blood cell and that the red blood cell had a higher water potential than the solution in the beaker.

**Time for a quiz**
- Review strategies in Chapter 2
- Take Quiz 3 at the REA Study Center
  (www.rea.com/studycenter)

# Chapter 11

# Homeostasis

## Homeostasis in Plants and Animals

**Homeostasis** is the maintenance of stable internal conditions in the body. It is generally controlled by a group of body sensors, the nervous system, and the endocrine system, which coordinately control several body systems. Body systems coordinate their activities (e.g., heart rate and respiration rate increase if the muscles need more oxygen or a plant begins to preserve its use of water during times of water limitation).

Each cell of a multicellular organism has the same genome, but cells differ from each other because they express different genes of the genome. The body of a multicellular animal is organized into groups of cells called tissues, groups of tissues called organs, and groups of organs called organ systems. Plants are multicellular, eukaryotic, autotrophic organisms organized into three main types of tissue systems—dermal, ground, and vascular. Each of the plant organs—roots, leaves, and stems—contains different structural arrangements of each tissue system.

The structure of a component of the body or plant system underlies its function.

## Coordinated Cooperation—Organelles

Macromolecules associate with each other to form the next level in the hierarchy of biological organization. This level is characterized by organelles, structures found within cells that carry out specific functions. The number and kinds of organelles differ from one kind of cell to another. Table 11-1 summarizes the functions of the major organelles. Individual organelles will be covered in detail as the process for which they are needed is discussed.

Table 11-1. Major Organelles of Cells

| Organelle | Function | Prokaryotic | Eukaryotic Cell | | | |
|---|---|---|---|---|---|---|
| | | | Protists | Fungi | Plants | Animals |
| Cell wall | Support and protection | ✓ | ✓ | | | |
| Plasma membrane | Control movement of cellular traffic | ✓ | ✓ | ✓ | ✓ | ✓ |
| Nucleus | Double lipid bilayer that contains DNA | | ✓ | ✓ | ✓ | ✓ |
| Nucleoid region | Contains single ring of bacterial DNA. Not separated from cytoplasm. | ✓ | | | | |
| Nucleolus | Assembles ribosomes | | ✓ | ✓ | ✓ | ✓ |
| Mitochondria | Double lipid bilayer with inner layer highly folded cristae and embedded with proteins of the ETC*. Production of ATP by oxidative phosphorylation occurs here. | | ✓ | ✓ | ✓ | ✓ |
| Chloroplast | Double lipid bilayer. Interior contains stacks of thylakoid bound grana embedded with chlorophyll and proteins of photosynthesis. Captures light energy and converts it to chemical energy. | | ✓ | | ✓ | |
| Ribosome | Composed of 2 subunits that translate mRNA into proteins | ✓ | ✓ | ✓ | ✓ | ✓ |
| Endoplasmic reticulum (ER) | Rough ER processes proteins for export. Smooth ER produces lipids. | | ✓ | ✓ | ✓ | ✓ |
| Lysosome | Intracellular digestion, apoptosis | | ✓ | ✓ | ✓ | ✓ |
| Vacuoles–large Vesicles–small | Vacuole: Large for support/storage in plants Vesicle: Small, formed by endomembrane system for transport | | ✓ | ✓ | ✓ | ✓ |
| Golgi apparatus | Final processing of proteins and lipids. Produces vesicles for export. | | ✓ | ✓ | ✓ | ✓ |
| Cytoskeletal elements | Intermediate filaments support and shape cells. Microfilaments and microtubules anchor proteins, move material. | | ✓ | ✓ | ✓ | ✓ |

* Electron transport chain

(*Continued*)

*(Continued)*

| Organelle | Function | Prokaryotic | Eukaryotic Cell | | | |
|---|---|---|---|---|---|---|
| | | | Protists | Fungi | Plants | Animals |
| Cilia/flagella | Move part or all of cell. In eukaryotes they have a 9+2 microtubule arrangement. | ✓ | ✓ | ✓ | ✓ | ✓ |
| Cell junctions | Communication between cells<br><br>Plasmodesmata in plants<br><br>Tight and gap junctions in animals | | | ✓ | ✓ | ✓ |

The cell's organelles interact with each other to make and use proteins, to obtain energy, and to carry out all the necessary metabolic functions of its characteristic cell type.

An example of organelle interaction can be seen when a cell produces a protein for export from the cell. The DNA is transcribed into mRNA. The mRNA moves into the cytoplasm where it associates with a ribosome. The ribosome begins translating the mRNA. The initial sequence of the polypeptide acts as a signal that this protein is destined for the endomembrane system or for secretion. With this signal polypeptide, proteins escort the ribosome to a receptor protein in the endoplasmic reticulum (ER) and translation is completed with the protein moving into the lumen of the ER. This ER dotted with ribosomes, is termed *rough* ER. Within the lumen of the ER, a protein will begin to assume its final conformation and move to the smooth ER, so called because there are no ribosomes present. A vesicle will bud off of the smooth ER and move to the Golgi apparatus (GA) where the protein will be stored and, potentially, further modified. A vesicle with the modified protein will bud off of the GA and move to fuse with the plasma membrane.

For a protein that is meant to be part of the endomembrane system or secreted from the cell, the efforts of at least five organelles and a myriad of supporting macromolecules working in concert is required. Figure 11-1 shows a schematic of the route followed by a protein with a signal sequence that indicates it should move to the rough endoplasmic reticulum for completion.

Figure 11-1. Route of a Protein Towards the ER

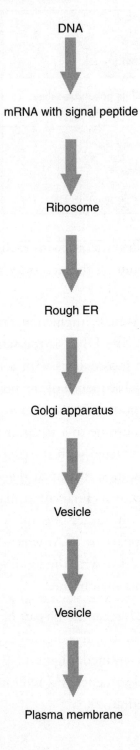

## Cells, Tissues, and Organs

Organelles come together to form cells. Groups of different specific cell types come together to form a tissue, which will accomplish a particular task. Groups of tissues come together spatially as an organ to perform a particular task, and organs work together as systems to support an organism. Cells, tissues, organs, and systems represent divisions of labor all designed to maintain a complex multicellular organism by maintaining a constant environment—sometimes called **homeostasis**—for every cell.

Cells consist of a plasma membrane and some or all of the cells contain organelles, depending on the cell's function. In the hierarchy of biological organization, cells are the first level of organization that can exhibit all of the characteristics of life. The cell theory states that the cell is the basic unit of structure and function of all living things and that all living things come from cells

All cells have a phospholipid plasma membrane embedded with proteins, DNA within a nucleus or in a nucleoid region, and cytoplasm. Cells are small to maximize the surface area to volume ratio. The smaller, narrower, or more folded a membrane is, the greater the surface area is present for exchange of materials between the cell and the environment.

Prokaryotic cells are smaller than eukaryotic cells, they do not have membrane-bound organelles, and their DNA is not separated from the cytoplasm. Many of the reactions that are performed by organelles in eukaryotes are accomplished by proteins embedded in the plasma membrane of prokaryotes. Prokaryotes do have ribosomes, though they differ from eukaryotic ribosomes. There are two kinds of prokaryotes, eubacteria and archaea. These two groups are so different from each other that scientists restructured the classification system to have three domains. A *domain* is taxonomically above a kingdom.

Eukaryotic cells have an extensive endomembrane system that subdivides the interior of the cell to allow for many different reactions to occur simultaneously. The nuclear envelope, the Golgi apparatus, the rough and smooth endoplasmic reticulum, and vesicles are all part of the endomembrane system.

A cell carries out its functions based on the kind of cell it is and the signals it receives. Most cellular responses are carried out by regulatory proteins that interact with DNA to initiate or stop transcription.

Most cells of multicellular eukaryotic organisms have their species' complete DNA in their nucleus even though the cells may be very different, such as skin cells and nerve cells. **Cell differentiation** occurs when a cell is committed to being a particular cell type and involves turning on some genes and turning off other genes. Once committed, a cell can only respond to specific signals.

Cell differentiation is an orderly process that begins before fertilization of the egg and is mediated at each step by regulatory proteins or by cell-to-cell signaling proteins. It occurs through a process of pattern formation, morphogenesis, and cellular differentiation.

Maternal contributions to the egg prior to fertilization begin the differentiation process. The maternal contributions, which are regulatory proteins and mRNA for regulatory proteins, are not uniformly distributed in the egg. The relative kinds and concentrations of these molecules determine the overall body plan. After fertilization, subsequent cell divisions form a ball of cells, the blastula. Each cell of the blastula does not contain the same proportion of the maternal contributions, and this difference sets the cell along its developmental path. For some organisms, a cell is committed to its final function immediately after the post-fertilization cell division. In other organisms, the cells at this point are still incapable of forming any cell.

The blastula reorganizes into layers through gastrulation; each layer of cells will become specific tissues and organs. Further differentiation occurs as a result of induction, regional changes due to signals sent by neighboring cells or through contact with the extracellular matrix of adjacent cells. **Morphogenesis**, the development of a body plan, then proceeds through the actions of inductive signals initiating cell movement, cell shape change, and cell adhesion. For example, coordinated limb bud lengthening occurs when growth factors secreted by one cell induces neighboring cells to enter mitosis. Once a cell is differentiated, all of its descendent cells retain the differentiation.

The process of cell differentiation can affect the cell's function. For example, cells that are destined to be cells of the lens of the eye activate the genes for crystallin, the fibers that will form the lens of an eye. Crystallin fibers are long, so as they are produced within a cell, they cause the cell to lengthen and flatten. This contributes to the shape and, ultimately, the optical properties of the lens.

Once differentiated and mature, a cell carries out its specific tasks in response to environmental changes. The environmental changes may mean changes in the fluid that bathes a cell—essentially its environment—or it may be changes from outside the organism. All organisms must make adjustments to optimize their survival chances. In plants, many of the adjustments are responses to changes in light availability, which is critical to plant cell metabolism. Response to light is governed by cells with specialized membrane structures called phytochromes.

**Phytochromes** are polypeptides that form as two identical subunits. Each subunit has a chromophore, a polypeptide sequence that has a pigment embedded and domains that extend into the cell membrane. A common phytochrome in plant cells is the Pr/Pfr phytochrome. Its chromophore can exist in two forms, depending on whether it is irradiated with light in the red part of the spectrum or the far-red part of the spectrum. In the Pr state, red light is absorbed and the chromophore converts to a Pfr absorbing form. This conversion is reversible and Pfr will revert to Pr if it absorbs a photon of far-red light or is kept in the dark. The biologically active form of the phytochrome is the Pfr form.

When Pr is illuminated with red light and isomerizes to Pfr, the shape of the membrane domain of the phytochrome changes. This embedded part of the phytochrome has kinase activity and sets off an intracellular response to light. The physiological response to the phytochrome depends on the tissue or organ in which the specialized cell is located. If the cell is in a seed, the conversion may set off activities for germination. In *Arabidopsis thaliana*, a small flowering plant, stimulation of phytochromes in leaves leads to activation of the flowering locus T gene. The protein product, florigen, acts as a long distance signal that travels to the shoot meristem. Here the florigen interacts with transcription factors to produce flowering proteins.

Close homologues of phytochromes have been found in photosynthesizing prokaryotes indicating a conserved element. This homology is another example of evolution modifying existing structures versus making brand new structures.

## Organ and Systems Interaction

Communication between organs and their systems is required to coordinate activities at each level. The components of this communication are sensory receptors, which receive a stimulus as some form of energy such as light or pressure, integrators that interpret the signal, and effectors, which respond to the signal.

Biological activities of complex organisms are dependent on coordinated interaction between organs. To illustrate, look at the relationship between the roots, stems, and leaves, three organs of plants.

The leaf is the organ of photosynthesis for a plant. It consists of epidermis, mesophyll, and vein tissue. The reactions of photosynthesis occur in the mesophyll tissue and require a constant supply of water and carbon dioxide. The root is the underground organ which acquires water and dissolved minerals from the environment to supply the leaf with the water for photosynthesis and other tissues of the plant with minerals and nutrients used in the synthesis of other molecules needed by the plant. The root can also store the excess products of photosynthesis. The stem consists of xylem, phloem, and epidermal tissues. The stem carries the raw materials from the roots to the leaves in xylem and the products of photosynthesis to all the living parts of the plant through phloem.

For a complex multicellular organism, the systems must coordinate interaction. Exchange of oxygen and carbon dioxide is common to many organisms because of the widespread metabolic use of oxygen as an electron acceptor in cellular respiration and the generation of carbon dioxide as a metabolic waste. This process is accomplished in different ways by different organisms. For example, plants rely on movement of gases, especially $CO_2$ through the stomata in leaves. Small flat animals rely on diffusion of the gases across their moist skin. In most animals, gas exchange occurs through the coordinated activities of the respiratory and circulatory systems.

The **respiratory system** extends from the mouth and nose to the lungs to enable $O_2$ to enter the body. Once the $O_2$ enters the body, the circulatory system carries the $O_2$ to the heart and out to all of the cells of the body through ever smaller vessels until each cell of the body is exposed to the blood carrying the $O_2$. The exchange of $O_2$ and $CO_2$ at the cellular level is carried out by diffusion. The $CO_2$ that is the waste product of cellular respiration diffuses into the blood cells and is carried through ever larger blood vessels back to the heart, the major organ of the circulatory system. The heart pumps the returned blood to the lungs for oxygenation. After oxygenation, the blood returns to the heart to repeat the gas exchange cycle.

The entire process is controlled by chemoreceptors in the medulla of the brain. These receptors are responding to a drop in the pH of the blood, which is indicative of too much $CO_2$ in the blood. The medulla sends signals to the muscles of the diaphragm to alter the rate of breathing. Breathing can also be altered when sensors in the heart detect a drop in blood pressure and send a signal to the brain.

## TEST TIP

All the systems of the body come together to maintain the homeostasis of the body. This is a central theme in the AP Biology curriculum.

## Negative and Positive Feedback Mechanisms

Organisms use **feedback mechanisms** to maintain their internal environments (that is, maintain homeostasis), respond to external stimuli, and regulate growth. Feedback does not work instantaneously; instead, it takes time to return to a steady state. Therefore, organisms take advantage and utilize both negative and positive feedback mechanisms.

**Negative feedback** occurs when a stimulus produces a result, and the result inhibits further stimulation. Negative feedback regulation is important in many physiological processes as varied as regulating gene expression, monitoring temperature, and releasing hormones. Its mechanism of action is that its disruptions to homeostasis cause a change in a stimulus, which activates a second stimulus that then decreases the original stimulus, resulting in the organism returning back to its original homeostatic state (see Figure 11-2). Essentially, process or substance A stimulates process or substance B. The presence of B then shuts off A. Not only does negative feedback help keep an organism's internal systems balanced, it also conserves energy by preventing unnecessary cellular reactions from occurring.

Figure 11-2. Negative Feedback Loop

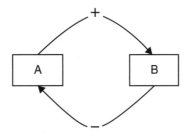

# TEST TIP

Negative feedback loops are a key concept to understanding homeostasis. It is a good idea to not only understand the basic mechanism as explained in Figure 11-2, but to also have an example ready to discuss in a free-response question.

There are many examples of negative feedback loops in biology, including:

- Operons in prokaryotic gene regulation. Many prokaryotes conserve energy by producing anabolic enzymes only when a substance is absent from the cell, such as in the *trp* operon. In this case, the absence of the amino acid tryptophan turns on the operon, which will cause the expression of genes for the enzymes that synthesize tryptophan. The subsequent accumulation of tryptophan will activate the repressor, causing the operon to turn off.

- Enzyme pathways. In cellular respiration, the production of ATP inhibits the function of the enzyme phosphofructokinase, a key enzyme in the catabolic breakdown of glucose in glycolysis.

- Thermoregulation in mammals. Mammals are **endothermic** and thus need to maintain a constant temperature. Mammalian thermoregulation is carried out by the **hypothalamus**, a region of the brain that acts as a thermostat. Essentially, the hypothalamus senses if body temperature varies from the norm and causes physiological changes (such as sweating) that will either increase or decrease the animal's temperature (see Figure 11-3).

Figure 11-3. Negative Feedback: Human Thermoregulation

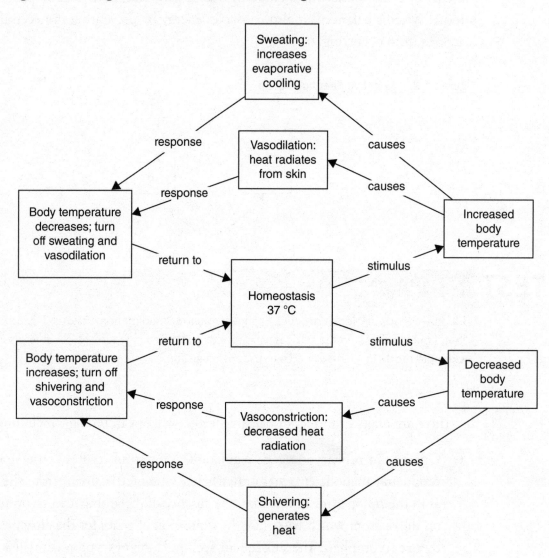

**Positive feedback** occurs when a stimulus produces a result, and the result causes further stimulation, thereby triggering an event. Positive feedback loops amplify a process in a physiological system. When stimulus A activates B, it causes B to activate A, which increases the response (see Figure 11-4). A common example occurs during labor in humans. During labor, the baby's head presses against the cervix. Pressure sensors located there cause the release of a hormone called oxytocin. This causes further contractions and more pressure on the cervix, releasing more oxytocin. This cycle increases the release of hormone and the rate of contractions, continuing until the baby is born and the pressure on the cervix is released.

Figure 11-4. Positive Feedback Loop

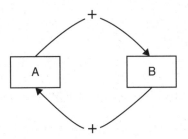

Any changes in a feedback mechanism can affect the health of an organism. One common example is Type 1 diabetes, which results from decreased insulin levels. In an unaffected individual, when blood glucose levels rise, beta cells in the pancreas release the hormone insulin. Increased concentrations of insulin in the blood trigger liver and body cells to remove glucose from the blood and store it (as glycogen if it is in the liver). This decreases blood sugar levels and returns the body to homeostasis. If blood sugar drops, then the hormone glucagon triggers the liver to break down glycogen and release glucose into the bloodstream. In a person with Type 1 diabetes, the pancreatic beta cells are destroyed by the immune system; therefore, insulin is not produced, causing spikes and drops in blood sugar levels and an inability to maintain homeostasis. If treated, Type 1 diabetes is a manageable disease; however, if left untreated, complications can include seizures, organ and tissue damage, coma, and eventual death.

> **DIDYOUKNOW?**
>
> Type 2 diabetes is the more common form of diabetes and works using a different mechanism, called insulin resistance. Essentially, a person's liver and body cells do not respond to increases in blood insulin levels. Thus, glucose isn't taken up into these cells for storage, leading to abnormally high blood glucose levels.

# Evolutionary Similarities and Examples

Homeostatic mechanisms are reflected in organisms of common ancestry. These mechanisms can continue to be similar in different organisms of different species over time or they can change in order to help organisms maintain homeostasis in their specific environments.

## Homeostatic Mechanisms and Continuity

Similar excretory systems are found in flatworms, earthworms, and vertebrates:

- Flatworms dispose of waste through tiny holes that are attached to internal tubes called protonephridia.

- Earthworms dispose of waste through tiny holes on their undersides that are attached to internal tubes called metanephridia.

- Vertebrates dispose of waste through two holes (anus and urethra) that are connected to tubes called nephrons within kidneys.

All three of these organisms demonstrate continuity in how they maintain homeostasis by disposing of waste from their bodies. The continuity in this specific homeostatic mechanism also demonstrates their common ancestry.

## Homeostatic Mechanisms that Change Over Time

Respiratory systems are found in both aquatic and terrestrial animals in the forms of gills (aquatic) and lungs (terrestrial). For the respiratory system to function, it requires a wet environment, large surface area, thin membranes, and the presence of $O_2$. In aquatic environments, gills are specifically customized to accommodate the organisms. Aquatic environments are extremely wet and have little $O_2$. Therefore, gills are on the outside of fish because they are able to be kept wet at all times by the surrounding water; they have a large surface area and thin membranes in order to effectively absorb as much $O_2$ as possible.

In terrestrial environments, lungs are specifically customized to accommodate the organisms. Terrestrial environments have little water, but lots of $O_2$. Therefore, organisms with lungs have them tucked safely inside their bodies in order to preserve their wetness. Lungs also have large surface areas and thin membranes in order to easily absorb $O_2$.

Although the same characteristics of the respiratory system exist for both aquatic and terrestrial organisms, over time, these organisms developed organs to help them "breathe" in their respective environments.

## Disruptions of Homeostasis

Biological systems are affected by disruptions in their dynamic homeostasis. These disruptions can occur at the molecular level, which affects the health of the organisms or at the level of the entire ecosystem, which affects the survival of parts or all of the population.

Foreign, toxic substances can cause molecular and cellular disruptions of homeostasis. As the substances enter the organism, they are detected by multiple cells, tissues, and organs that try to expel them from the organism and to reinstate homeostasis. The body combats these disruptions through physiological and immunological responses.

Disruptions to an ecosystem's homeostasis can come from a variety of sources. Other species can disrupt the homeostasis of another species through predation or parasitism. Humans can impact not only their ecosystems, but also those of other species; an example would include contamination of a local lake that is not only a water source for humans, but also home to fish, insects, birds, etc., that are also affected by the contamination. Hurricanes, earthquakes, or floods can also cause disruptions for ecosystems. Also, the limitation of water can cause serious problems for any organism and threaten its internal homeostasis.

Because organisms need to keep their internal systems constant, any changes in their external environment will cause the organism to respond in an attempt to stay at homeostasis. These responses can be either behavioral or physiological and include the following three physiological examples:

1. **Photoperiodism** in plants. The photoperiod is the relative amount of light during the day and night; summer has longer days than winter and thus more light is available for plants to use for

> **DID YOU KNOW?**
> Poinsettias, the red and green plants that are commonly used as decoration at Christmas time, are a common example of a short-day plant that flowers when it receives less light. Many people try to save their poinsettias after the holidays and hope that they will turn red again. However, few people accomplish this because they provide the plants with too much light for them to flower again.

photosynthesis. Plants need to conserve resources during the winter because less light means they do not produce as much energy through photosynthesis. For example, deciduous trees respond to the shortening of days in autumn by dropping their leaves, ultimately preventing the loss of water through transpiration. Additionally, plants need to sense when light is more available in spring so they can produce leaves to photosynthesize and flower to reproduce.

2. **Hibernation** in animals. Many mammals, living in habitats that have cold winters, enter into a state of dormancy in order to conserve energy. In addition to the change in temperatures, food is typically scarce. Thus, an animal that responds to winter changes is more likely to survive than one who attempts to stay warm and find food. During hibernation, the mammal's core body temperature drops and its metabolic rate decreases. Ultimately, the animal survives on stores of fat it accumulated during the summer or on food it hid away in its burrow.

3. **Shivering and sweating** in humans. Shivering is a type of thermogenesis that generates heat when the external environmental temperature has dropped, making the maintenance of homeostasis difficult. Shivering occurs when the muscles (especially those around the core) contract rapidly, requiring an increase in cellular respiration. The heat generated through metabolism is used to warm the body. Sweating occurs when the environmental temperature has increased and body temperature has risen above the stable state. Sweat glands on the surface of the body release sweat, which cools the body as it evaporates (one of the properties of water), thus returning the body to a lower temperature.

> **DID YOU KNOW?**
> Shivering can generate up to ten times the amount of heat that is produced when the body is at homeostasis.

## Questions to Consider

1. Make a table comparing and contrasting negative and positive feedback loops.

2. Because of global warming, temperatures during late winter to early spring are often higher than they have been historically. Some animals, such as brown bears, are waking out of hibernation earlier in the year than they do when temperatures are normal. Explain how this disruption is leading to the death of some bears.

3. Suppose you are a genetic counselor and Isaac and Khalilah, a young couple wanting to start a family, comes to you for advice. Isaac has a large family and a number of his siblings (both sexes) suffer from mitochondrial encephalomyopathy, a disease caused by a mutation in mitochondrial DNA. Khalilah has no such history in her family. What would you tell this young couple about the prospects for this disease showing up in any children they might have?

## Answers:

1. 

| Negative Feedback | Both | Positive Feedback |
|---|---|---|
| —Dampens a response<br>—Stimulus decreases activity<br>—Create less of a product | —Physiological mechanism<br>—Involved in homeostasis<br>—Stimulus activates the loop | —Amplifies a response<br>—Stimulus increases activity<br>—Create more of a product |

2. Animals hibernate in order to conserve energy, especially because their habitat lacks food sources during the winter. If a bear wakes up too early, the plant and animal species it eats will not be available yet, and the bear is likely to die of starvation.

3. You would explain to Isaac and Khalilah that this particular mitochondrial disease is maternally inherited. That is, the mutation exists in Isaac's mother's mitochondrial DNA. Nearly all of a child's mitochondria come from the mother because she forms the egg and supplies it with most of its initial organelles, such as the mitochondria. The sperm contributes only nuclear DNA to offspring inheritance. Since Khalilah's family has no history of this problem, this couple is likely to have normal children with regard to mitochondrial inheritance. Isaac will not be contributing his mitochondria to his offspring.

# Chapter 12

# Reproduction, Growth, and Development

## Reproductive Process of All Land Plants

The **alternation of generations** is the sexual life cycle of all plants, although sexual reproduction specifically varies from group to group. Unlike animals in which the diploid organism is the only multicellular form, in plants there are two multicellular forms: the *gametophyte* and the *sporophyte*. A gametophyte is a haploid and produces the egg and sperm; the sporophyte is diploid and is formed by the fusion of egg and sperm. A sporophyte also produces haploid spores by meiosis and is the product of fertilization of male and female gametophytes.

Figure 12-1. Alternation of Generations

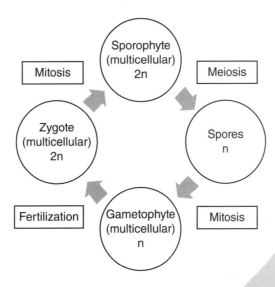

# Plant Growth and Development

Modifying gene expression can affect the timing of development. For example, in plants, a seed must time its development with water availability or it will be unable to grow and photosynthesize. **Germination** is the process in which a plant embryo emerges from a cell and begins growing; water is one trigger of germination. Seeds have low **water potential**; thus, if water is available, the seed will absorb water, which causes the release of plant hormones called gibberellins. Gibberellins set off a signal cascade that triggers the transcription of an enzyme (alpha-amylase), which will digest the starch stored in the seed. Because the plant embryo does not have leaves to photosynthesize, this starch acts as the source of energy, allowing the young plant to grow its first roots, stems, and leaves.

There are numerous environmental factors, such as drought, extreme temperatures, or lack of sufficient nutrients available in the soil, that can negatively affect the growth and development of plants. If these adverse conditions are brief in time and not too extreme, then chances are that the plant will survive, although there might be leaf loss or diminishment to the overall health of the plant.

How plants respond to adverse environmental conditions:

- In drought situations, plants often close the stomata in order to reduce transpiration (loss of water through the plant's leaves). The purpose of transpiration is to cool the plant by pulling water and nutrients from the soil to parts of the plant, especially the leaves. Extended drought conditions cause the plant's water supply to diminish, which inhibits the plant from regulating its temperature. Ultimately, the plant can become nutrient-deficient and photosynthesis can be compromised.

- In extreme temperatures, the plant conserves its water supply and increases transpiration in order to help stabilize its temperature; if more water is unavailable, then extreme temperatures can negatively affect the health of the plant.

- Leaf loss allows the plant to minimize the areas in which water must be dispersed in order to maintain the plant's overall temperature. In extreme heat or drought, leaf loss lets the plant to "cope" and attempt to preserve itself.

- A genetic mutation is not always a bad thing for a plant; however, any mutation that compromises the plant's ability to obtain nutrients, to grow and develop, and to maintain homeostasis can have an adverse effect on the plant and possibly result in death.

# Reproduction

## Reproductive Patterns

**Asexual reproduction** allows for no genetic diversity since all genes come from one parent. There is no fusion of egg and sperm. **Buddings**, which are outgrowths from the parent, form and pinch off to live independently. **Binary fission** is a type of cell division by which prokaryotes reproduce. Each daughter cell receives a single parental chromosome. **Fragmentation** is the breaking of a body piece that will form an adult via regeneration of body parts.

**Sexual reproduction** provides more genetic diversity since genes will be inherited from both parents. It involves the fusion of an egg and a sperm.

In **parthenogenesis**, the egg develops without being fertilized and in **hermaphroditism**, the organism has both male and female reproductive organs.

**Spermatogenesis** is the production of sperm and is stimulated by the hormone testosterone. It occurs in the seminiferous tubules of the testes, which are located in the scrotum. Spermatogenesis is continuous throughout the life of a male. During each meiosis, four viable sperm are produced, yet, meiosis occurs in an uninterrupted sequence, unlike in females.

**Oogenesis** is the production of ova and happens in the ovaries. Egg cells begin with meiosis, but are arrested in *prophase I* until one egg per menstrual cycle is stimulated to complete meiosis by the hormone *FSH*. Meiosis then stops again and the ova does not undergo *meiosis II* until after fertilization. At birth, the ovary contains all the cells that will develop into eggs from puberty to menopause. Only one viable ovum is produced at a time with three nonviable polar bodies.

The reproductive cycle of the human female, also called the menstrual cycle, creates changes in the uterus or female reproductive organ. It is a 28-day cycle in which the destruction and regeneration of the uterine lining (endometrium) occurs. There are three phases of the human female reproductive cycle:

1. **Menstrual Phase** (day 0–5)—menstruation (bleeding due to destruction of endometrium).

2. **Proliferative Phase** (day 6–14)—regeneration of endometrium.

3. **Secretory Phase** (day 15–28)—endometrium becomes more vascularized and is ready for implantation of embryo. If the embryo is not implanted, the entire menstrual cycle will happen again.

The **ovarian cycle** parallels the menstrual cycle. The ovarian cycle undergoes the follicular phase (day 0–13) during which the egg cell enlarges in a follicle. During ovulation (day 14), the oocyte is released and pregnancy can take place. In the luteal phase (days 15–30), the corpus luteum is formed, which is a structure that grows on the surface of the ovary where a mature egg was released at ovulation. The corpus luteum produces progesterone, which prepares the body for pregnancy.

Hormones play a very important role in the menstrual cycle. Under the control of four endocrine hormones—*LH, FSH, estrogen,* and *progesterone*—one of two ovaries releases one egg cell per menstrual cycle. *LH* (luteinizing hormone) and *FSH* (follicle stimulating hormone) are made in the anterior pituitary. Estrogen and progesterone are made in the ovaries. *FSH* stimulates ovulation to occur. Estrogen causes cell division in the uterine lining and stimulates the release of LH during menstruation; during the luteal phase, estrogen causes LH and FSH levels to fall. When LH levels spike during menstruation, the follicle bursts, releasing the egg and progesterone causes blood vessel growth in the uterine lining during the luteal phase.

## Animal Development

**Fertilization** is the process of the egg and sperm fusing to make a zygote.

The structure of sperm is important to how they function. The head of a sperm cell contains the nucleus carrying its 23 chromosomes. At the front of the head is an area containing digestive enzymes that are used to digest the layer of cells surrounding the egg. Immediately below the head is the midpiece, which contains the many mitochondria it uses to make ATP to power its flagellum. Each sperm has a single flagellum, which moves to propel the sperm forward.

Figure 12-2. Structure of a Sperm Cell

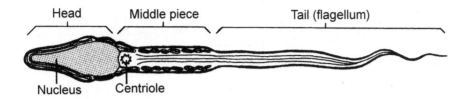

An ejaculation produces around 300 to 400 million sperm, but most do not survive.

When sperm contact the egg, they do so with their heads, and they release their enzymes. When one sperm reaches the egg's plasma membrane, it fuses with it. The flagellum detaches, and only the head of the sperm and its midpiece enter the egg. Most of the time, only one sperm enters an egg because when it fuses with the egg, then the egg rapidly blocks any remaining sperm from entering.

Fertilization usually occurs in the fallopian tubes of the female reproductive tract. The fertilized egg then begins to divide by mitosis and moves to the uterus where it attaches to the uterine lining and finishes embryonic and fetal development.

Figure 12-3. Unfertilized Egg Cell

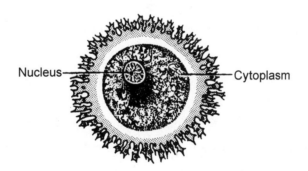

After fertilization of an egg and formation of a zygote, timing in the earliest phases of development is crucial for organisms to function normally. One of the first steps in animal development is **differentiation**, which occurs when a cell becomes specialized for a specific process, such as a muscle cell. As a zygote divides in a process called **cleavage**, it becomes a hollow ball of cells called a **blastula**.

As the blastula continues to divide, it will fold in on itself during gastrulation and form three layers: the ectoderm (outermost), the mesoderm (middle), and the endoderm (innermost). Once the layers have formed, cell differentiation occurs as individual cells receive inductive signals that determine cell fate. Inductive signals are responsible for switching on or off certain genes. Cells of the ectoderm are destined to become epidermal and nervous tissues. Mesodermal cells will differentiate into skeletal, muscular, and circulatory/lymphatic cells. The linings of the digestive, respiratory, excretory, and reproductive tracts are derived from the endoderm.

> **DID YOU KNOW?**
> Embryonic stem cells are taken from the cells of a blastula. They are still able to differentiate into many different types of cells, which is why they are of great interest to medical science.

> **TEST TIP**
> It is not necessary to memorize the stages of embryonic development, such as gastrulation, or the specific layers and their specialized tissues. You should understand that during development inductive signals caused by gene expression cause cells to differentiate and develop specific functions.

An example of inductive signaling occurs in the development of limbs in vertebrate animals. Developing vertebrates have a thickened region in the epidermis of their limb buds called the **apical ectodermal ridge**. Cells in this ridge secrete several proteins, most notably fibroblast growth factors, to the cells around them. These growth factors regulate gene expression of the cells in the limb and promote proper growth of vertebrate limbs.

Many of the inductive signals regulating cell differentiation set off signaling cascades. The result of the cascade is to trigger transcription factors that regulate gene expression or affect mRNA post-transcriptionally. The following is an example of ways modifying gene expression affects the timing of development:

Homeotic genes control the spatial organization in the embryonic development of plants and animals because their protein products act as transcription factors. In animals, they are called *Hox* genes and are such an important group of genes that they are highly conserved, meaning that either their sequences are nearly identical or that they are spatially arranged similarly (meaning the gene for the head is located before the gene for the body). Typically, *Hox* genes are master regulators that bind to many regions of DNA and coordinate the transcription of multiple genes required for animal development. Mutations in homeotic genes can cause abnormal development. In one study of mice having two mutant forms of *Hox* genes, the forelimb was misshapen because the humerus, radius, and ulna (the three main bones of the forelimb) formed improperly.

Small single-stranded RNA molecules, called microRNAs (miRNAs), regulate genes post-transcriptionally and help control cellular functions and development. miRNAs form complexes with a class of proteins called argonautes. The complexes they form are called RISCs (RNA-induced silencing complexes), and they bind to strands of mRNA having a complementary base sequence. The RISCs either interfere with the translation of mRNA by blocking its access or degrade the mRNA completely, both of which prevent the synthesis of the gene's protein product.

> **DID YOU KNOW?**
>
> miRNAs were only discovered in the 1990s, and their regulatory properties weren't understood until the 2000s. In this short period of time, scientists have come to understand that they are of great importance in gene regulation. For example, it has been estimated that half of human genes are regulated by miRNAs.

As organisms continue to develop, programmed cell death (apoptosis) occurs. Although "death" might sound bad, this is a very normal part of the development process that eliminates unnecessary cells. Apoptosis is not regulated through gene expression; instead, it consists of a pathway that activates proteins that are always found in cells. The cellular machinery responsible for cell death was first understood through studies with *C. elegans* (a type of worm whose entire genome is mapped). Scientists discovered that three genes control apoptosis: ced-3, ced-4, and ced-9. The proteins ced-3 and ced-4 are typically in an inactive form. However, when ced-9 receives a death signal, it triggers the activation of ced-3 and ced-4, which in turn activate proteases (enzymes that destroy proteins) and nucleases (enzymes that destroy DNA) that will destroy the cell. In mammals, there are several signaling cascades that activate proteases and nucleases; however, the overall process is similar to that studied in *C. elegans*.

Apoptosis occurs because, through evolution, developmental genes have been highly conserved. Thus, cells that might be needed in one organism are not needed in another. One common example is that many animals have genes that regulate the development of skin, called webbing, between digits. In most birds and mammals, however, this webbing is eliminated through apoptosis (ducks are an example of an animal that maintains the webbing).

> **DID YOU KNOW?**
>
> Humans are occasionally born with some webbing between their fingers or toes. This is called *syndactyly* and is caused by one of several different genetic disorders that prevents apoptosis from occurring properly during the sixth to eighth week of fetal development.

## Question to Consider

1. Scientists studying disease are working with stem cells because they hypothesize stem cells can be transplanted into patients with various diseases such as Parkinson's disease and the stem cells will replace the damaged cells. How will this happen?

## Answer:

1. Stem cells have not differentiated yet and are capable of becoming specialized cells. If a stem cell is transplanted into tissues, scientists hope that local inductive messengers will act on the still plastic stem cells. The messengers would signal the stem cells to differentiate into the tissues before they were damaged, providing a replacement for the nonworking cells.

# AP Biology Labs: Biological Processes and Energetics

Chapter 13

## Lab 4: Diffusion and Osmosis

The transport of materials across cell membranes and throughout cells is vital for cells to maintain homeostasis. The movement of materials from an area of high concentration to low concentration is **diffusion**. Diffusion of small particles can occur across the phospholipid bilayer of the cell membrane while larger particles move through membrane proteins. The diffusion of water across a selectively permeable membrane is called **osmosis**. It occurs across the cell membrane, but is fastest moving through membrane proteins called aquaporins. Both diffusion and osmosis are passive processes that do not require energy input.

Like diffusion, water moves from an area of high **water potential** (high water and low solute) to an area of low water potential (low water and high solute). Solutions that are separated by selectively permeable membranes are named for the amounts of solute in the solution. An animal cell placed into one of these solutions may gain or lose water, making it a challenge to maintain homeostasis.

The table below summarizes the three types of solutions in comparison to the cytosol of an animal cell as well as what happens to an animal cell placed into one of these solutions.

| Solution | Amount of solute compared to cell | Water potential | Water movement of a cell in this solution | Effect on animal cell |
|---|---|---|---|---|
| Hypertonic | Higher | Lower | Out of cell Into hypertonic solution | Cell shrivels |
| Hypotonic | Lower | Higher | Into cell Out of hypotonic solution | Cell bursts |
| Isotonic | Equal | Equal | No movement | Cell normal |

Understanding water potential in plant cells is more difficult because of their cell walls. The cell wall exerts pressure on the cellular contents. This prevents the cell from bursting when placed in a hypotonic solution. **Turgor pressure** is the pressure of the cell membrane against the cell wall in a plant cell that is full of water; it keeps a plant cell from bursting when it is placed into a hypotonic solution and is necessary for plants to maintain rigidity.

When considering water potential in a walled cell, two factors must be considered: solute concentration (called solute potential) and the pressure exerted on the solution (called pressure potential). The Greek letter psi ($\Psi$) is used to represent water potential, so mathematically this is summarized as:

$$\text{water potential} = \text{solute potential } (\Psi_S) + \text{pressure potential } (\Psi_P)$$

or

$$\Psi = \Psi_S + \Psi_P$$

Increasing the solute will decrease water potential, while increasing the pressure will increase the water potential. This is because solute potential is a negative number whenever there is solute present. The formula is:

$$\Psi_S = -iCRT$$

i = the ionization constant of the solute

C = the molar concentration of the solution

R = the pressure constant (0.0831 liter bars/mole-K)

T = temperature in K (273 + °C).

For more explanation and examples of this concept, refer to your lab manual.

## Science Practices

There were three components to this lab. You may have done all three or just one, but each involved a slightly different concept.

1. Surface area to volume calculations using an artificial cell.

   In this activity, you cut out small cubes of agar that were a pink color due to the presence of an indicator (phenolphthalein) that is pink in a basic solution. You calculated the surface area and volume of each cube and placed them into a weak acid to measure diffusion. The acid caused the agar to change from pink to colorless. After several minutes, you measured the depth of color change in the pink blocks.

2. Creating cell models with dialysis tubing to investigate solutions and osmosis.

   For this portion of the lab, you were given solutions of 1 M sucrose, 1 M NaCl, 1 M glucose, and 5% egg white protein. The procedure involved setting up 5 different combinations of one solution in a dialysis bag, and the bag was placed into a beaker of another solution. The combination of solutions was determined by your group. For example, you might have placed 1 M sucrose inside the dialysis tubing and then placed it into 1 M NaCl. Your control should have been distilled water inside the tubing placed into a beaker of distilled water. You measured diffusion by massing the dialysis tubes before and after you placed them into the solutions.

3. Using living plant cells to observe osmosis.

   You designed an experiment to measure water potential in plant cells using sucrose solutions of varying molarity (0, 0.2, 0.4, 0.6, 0.8, and 1.0 M). The first thing you might have had to do was determine the concentrations of the solutions. This was accomplished by putting distilled water into dialysis tubing and placing it into each of the unknown solutions. The bag that lost the most mass was in the 1.0 M solution, and the bag that stayed the same was in the 0 M solution. Once you determined the unknown solutions, you used them with cores of some vegetable, such as potato or sweet potato. You massed the vegetable pieces before they went into the solution and again after a certain amount of time passed. Then you calculated the percent change in mass.

## Data Analysis

1. Surface area to volume calculations using an artificial cell.

    Below is a table showing data for cubes of 1 cm³, 2 cm³, and 3 cm³. Even if your cube sizes were different than these, your largest cube had the smallest surface area to volume ratio. The vinegar was not able to diffuse all the way through the largest cube.

| Cell | Dimensions (cm) | Surface Area (cm²) | Volume (cm³) | SA to Volume Ratio | Depth of Diffusion |
|---|---|---|---|---|---|
| 1 | 1 cm | 6 | 1 | 6:1 | Entire cube |
| 2 | 2 cm | 24 | 8 | 3:1 | Entire cube |
| 3 | 3 cm | 54 | 27 | 2:1 | 1.7 cm |

2. Creating cell models with dialysis tubing to investigate solutions and osmosis.

    For this portion of the lab, data is presented for a dialysis bag of each solution placed into a beaker of distilled water. If you calculate the solute potentials of each solution, you can estimate which way water would have moved. In the tubing, pressure potential was 0 because it was an open system, thus water potential equaled solute potential (using the formula given in the introduction). For these calculations, a temperature of 22 °C was used. One thing you might have noticed is that egg white albumin had little effect. Another thing to realize is that both NaCl and glucose are small enough to fit through the pores of the dialysis bag, so while water was diffusing in, solute particles were also diffusing out.

| Solution | Solute Potential | Dialysis Bag Mass Change |
|---|---|---|
| 1 M sucrose | −1(1M)(0.0831 liter bars/mole-K)(295K) = −24.5 bars | Gained |
| 1 M glucose | −1(1M)(0.0831 liter bars/mole-K)(295K) = −24.5 bars | Gained |
| 1 M NaCl | −2(1M)(0.0831 liter bars/mole-K)(295K) = −49.0 bars | Gained |
| 5% albumin | −1(0.001)(0.0831 liter bars/mole-K)(295K) = 0.02 bars | Equal or gained little |

3. Using living plant cells to observe osmosis.

   Data and a graph for the percent change in mass of several vegetables are shown below.

| Solution | Potato | Sweet Potato | Turnip | Beet |
|---|---|---|---|---|
| Distilled water | 15.306 | 17.2 | 28.13 | 15.03 |
| 0.2 M sucrose | 2.608 | 10.9 | 11.76 | 6.785 |
| 0.4 M sucrose | −15.034 | 6.87 | 7.25 | 0.58 |
| 0.6 M sucrose | −25.38 | 2.205 | 2.94 | −17.94 |
| 0.8 M sucrose | −36.002 | −3.235 | −9.09 | −27.13 |
| 1.0 M sucrose | −38.656 | −3.615 | −17.39 | −35.635 |

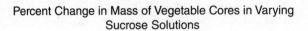
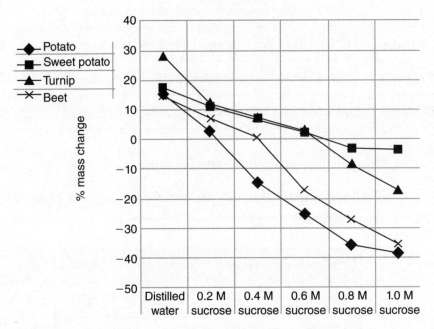

The graph is used to calculate the solute potential of each vegetable. The point where the line crosses the x-axis is the molarity of the sucrose solution that is equal to the water potential of the vegetable (because pressure potential equals zero). Calculations for each of the vegetables at 22°C using $\Psi_s = -iCRT$ are below.

| Vegetable | Line Crosses x-axis | Water Potential |
|---|---|---|
| Potato | 0.22 | −1(0.22M)(0.0831 liter bars/mole-K)(295K) = −5.4 bars |
| Sweet Potato | 0.69 | −1(0.69M)(0.0831 liter bars/mole-K)(295K) = −16.9 bars |
| Beet | 0.40 | −1(0.40M)(0.0831 liter bars/mole-K)(295K) = −9.8 bars |
| Turnip | 0.67 | −1(0.67M)(0.0831 liter bars/mole-K)(295K) = −16.4 bars |

## Conclusions

In the first part of this lab, it was clear that diffusion did not proceed as far in a "cell" having a larger volume. Thus, one of the limits to cell size is the surface area to volume ratio because it would take too long to move materials throughout a large cell.

In part two, movement of water was higher when there was a great difference in water potential between two solutions. This is why putting salt onto slugs or snails kills them or why de-icing salt used in the winter damages road-side plants; there is a very high difference in the water potential of pure salt and the cells of these living organisms. The reason there was little effect on the albumin is because it is a very large protein and won't diffuse through the membrane. Additionally, a 5% solution of albumin has a molarity of only 0.001. Thus, it is only slightly hypertonic to water and would show very little change in mass when placed into water.

In part three, water potential was calculated for different vegetables using percent mass change in each vegetable. Each vegetable showed a different water potential which is because each of these vegetables stores different amounts of sugars in the plant, giving each of them differing solute concentrations.

## Lab 5: Photosynthesis

Plants are **autotrophic** organisms capable of converting light energy into chemical energy through the process of photosynthesis. The general equation is:

$$2\ H_2O + CO_2 + light \rightarrow CH_2O\ (carbohydrate) + O_2 + H_2O$$

In order to determine the rate of photosynthesis, you can measure either the production of $O_2$ or the consumption of $CO_2$; this lab used oxygen production.

### Science Practices

You designed an experiment to test the effects of different conditions on the rate of photosynthesis. This lab used a simple procedure called the floating leaf disk method to measure the production of oxygen in order to determine the photosynthesis rate. A leaf consists of layers of photosynthetic tissues, vascular tissue, and air spaces.

First, you used a hole punch to cut out small leaf disks. Next, you used a syringe filled with sodium bicarbonate to remove all gases from the air spaces in the disks and filled them with sodium bicarbonate. Then you placed the leaf disks into a solution of sodium bicarbonate (a source of carbon dioxide); they sank to the bottom because of the lack of air in them. Once your experiment was set up to test the variable, you used a light to allow photosynthesis to proceed; the oxygen produced accumulated in the air spaces and caused the leaf disks to float.

## Data Analysis

The number of floating leaf disks were counted over time. A simple graph of this data for photosynthesizing leaf disks would look like the following:

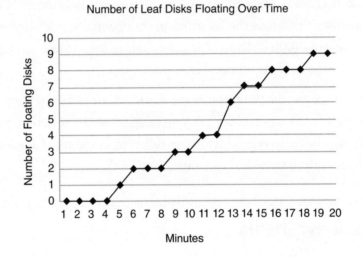

In your variable-testing experiment, you calculated rate of photosynthesis as the number of disks floating per minute. Essentially, the speed at which the leaf disks rise is correlated to the photosynthetic rate. For example, a graph showing how the photosynthesis rate varied with different concentrations of sodium bicarbonate is given.

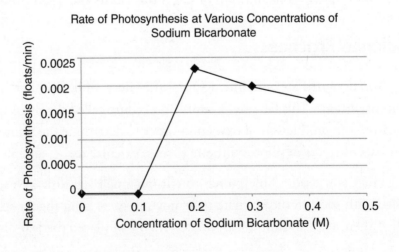

## Conclusions

The rate of photosynthesis varied depending upon the conditions tested. Some of the variables that increased the photosynthesis rate were increased light intensity, increased sodium bicarbonate concentration, near-neutral pH, and a temperature around 25–35 °C.

# Lab 6: Respiration

The molecule that cells use to carry out life processes, including growth, reproduction, and maintenance of homeostasis, is ATP. Autotrophs begin this process by making chemical energy from light energy, usually through photosynthesis. Both autotrophs and heterotrophs then convert chemical energy from carbohydrates, typically glucose, to ATP. This process is called **cellular respiration**. Respiration can be either **aerobic** or **anaerobic**. This lab focused on the aerobic process, which produces a greater amount of energy from each glucose molecule. The formula for aerobic cellular respiration is:

$$C_6H_{12}O_6 + 6O_2 \rightarrow 6CO_2 + H_2O + ATP$$

There are three ways to measure the rate of cellular respiration: the consumption of oxygen, the production of carbon dioxide, or the release of heat energy.

## Science Practices

This lab used a simple device called a respirometer to measure the consumption of oxygen to determine the rate of respiration. As oxygen is consumed in respiration, carbon dioxide is produced, so potassium hydroxide (KOH) was used to remove the carbon dioxide. KOH reacts with $CO_2$ to form a solid, potassium carbonate ($K_2CO_3$). The balanced equation is:

$$CO_2 + 2KOH \rightarrow K_2CO_3 + H_2O$$

By removing carbon dioxide gas from the system, as oxygen was consumed in cellular respiration, the volume of gas in the respirometer decreased.

Exact construction of respirometers depended upon your teacher, but the basic set-up consisted of a syringe or other tube filled with a living organism (such as germinating seeds or a small insect) or glass beads (the control), nonabsorbent cotton to shield the living organism from the KOH, and cotton covered in KOH. The respirometer was either attached to a calibrated pipette so you could directly measure the volume of oxygen used or it was attached to a gas pressure sensor so you could measure the change in air pressure. A control respirometer filled with an equal volume of beads was used because each respirometer is sensitive to changes in ambient air temperature and pressure. Once constructed, you designed an experiment to test the effect of some variable on respiration or to compare respiration rates in different organisms. For example, you might have compared rates in different types of seeds or insects, or examined the effect of temperature on respiration rate.

## Data Analysis

There were many possible experiments you could design; however, this book will review a comparison of three types of germinating seeds (English peas, black-eyed peas, and navy beans) at two different temperatures (10 °C and 25 °C). The data presented and graphed here is for a respirometer connected to a pipette; thus, the respiration rate was measured in mL of oxygen consumed over time. One important aspect of data analysis was to calculate a corrected difference by subtracting any changes that occurred in the control respirometer. Including the corrected difference in the rate insured that changes in the respirometer were due to oxygen consumption only. In the table below, "distance" is measured as the mL mark on the pipette at that time; "differ" is the difference between the two times; and "corrected differ" is the difference of the seeds with the bead difference subtracted. (The corrected difference is highlighted because that is the data used to construct the graph.)

| Temp | Time min. | Beads | | English peas | | | Black-eyed peas | | | Navy beans | | |
|---|---|---|---|---|---|---|---|---|---|---|---|---|
| | | Distance | Differ | Distance | Differ | Corrected Differ | Distance | Differ | Corrected Differ | Distance | Differ | Corrected Differ |
| 10 °C | 0 | 0.90 | | 0.9 | | | 0.9 | | | 0.9 | | |
| | 5 | 0.88 | 0.02 | 0.86 | 0.04 | **0.02** | 0.84 | 0.06 | **0.04** | 0.81 | 0.09 | **0.07** |
| | 10 | 0.87 | 0.03 | 0.75 | 0.15 | **0.12** | 0.73 | 0.17 | **0.14** | 0.70 | 0.20 | **0.17** |
| | 15 | 0.88 | 0.02 | 0.73 | 0.17 | **0.15** | 0.70 | 0.20 | **0.18** | 0.67 | 0.23 | **0.21** |
| | 20 | 0.87 | 0.03 | 0.68 | 0.22 | **0.19** | 0.65 | 0.25 | **0.22** | 0.63 | 0.27 | **0.24** |
| 25 °C | 0 | 0.90 | | 0.9 | | | 0.9 | | | 0.9 | | |
| | 5 | 0.89 | 0.01 | 0.78 | 0.12 | **0.11** | 0.74 | 0.16 | **0.15** | 0.76 | 0.14 | **0.13** |
| | 10 | 0.88 | 0.02 | 0.69 | 0.21 | **0.19** | 0.62 | 0.28 | **0.26** | 0.65 | 0.25 | **0.23** |
| | 15 | 0.86 | 0.04 | 0.59 | 0.31 | **0.27** | 0.57 | 0.33 | **0.29** | 0.54 | 0.36 | **0.32** |
| | 20 | 0.89 | 0.01 | 0.49 | 0.41 | **0.40** | 0.44 | 0.46 | **0.45** | 0.42 | 0.48 | **0.47** |

From the graph, it is clear that all seeds respired at a faster rate at room temperature than at 10 °C. Additionally, if you calculate the rate of respiration at 25 °C (corrected difference/time) the navy beans respired at a rate of 0.025 mL/minute, the black-eyed peas at a rate of 0.023 mL/minute, and the English peas at a rate of 0.020 mL/minute. Thus, the navy beans respired at a slightly higher rate than the other seeds at 25 °C.

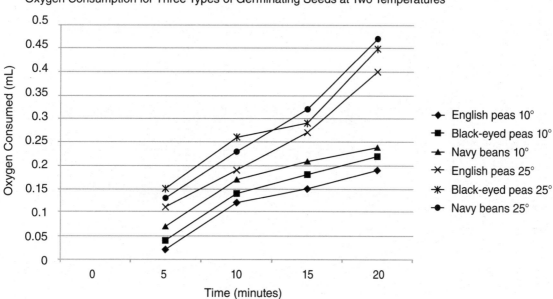

## Conclusions

As with other systems, in cellular respiration environmental factors affect the rate of the reaction as living organisms work to maintain homeostasis. Typically in plants, the higher the temperature, the faster the rate of respiration (assuming you didn't use a temperature that harmed or killed the seeds). This was true for the three types of germinating seeds; although there wasn't a great difference in their individual rates of respiration, all seeds respired at a greater rate at 25 °C than at 10 °C. However, this isn't true for animals because you must take into account whether the animal is warm- or cold-blooded and how hard it has to work to maintain its homeostatic temperature.

For example, follow-up questions for this lab often involve comparing different animals' respiration rates at varying temperatures. A mammal would have a higher respiration rate at a cold temperature than a reptile because it is warm-blooded and must expend more energy to maintain homeostasis. Similarly, if you compared the mammal's respiration rate at 25 °C versus 10 °C, it would have a higher rate at 10 °C for the same reason.

**Time for a quiz**
- Review strategies in Chapter 2
- Take Quiz 4 at the REA Study Center
  *(www.rea.com/studycenter)*

**Take Mini-Test 1**
on Chapters 3–13
Go to the REA Study Center
*(www.rea.com/studycenter)*

# DNA Structure and Replication and RNA Structure and Gene Expression

## Chapter 14

### Key Concepts

Living things respond to their environment, grow, and reproduce. These abilities are due to instructions carried on the molecule named DNA (deoxyribonucleic acid). The diversity of life forms, even the differences between members of the same species, is due to the differences in DNA from one organism to another. DNA functions by directing the production of proteins, which ultimately carry out the work of life.

Three important characteristics necessary for a molecule to qualify as the genetic material are that it must be able to be replicated accurately, it must be stable, and it must be able to carry a lot of information.

The structure of a DNA molecule is the key to understanding how each strand of DNA can act as a template for the replication of the other strand during DNA replication and for the production of RNA during transcription.

## Discovery of DNA as the Genetic Material

Initially, scientists hypothesized that proteins would be the genetic material because of the diversity and complex structure of proteins. A substantial body of experimental evidence was gathered that determined that, DNA—not proteins—is the genetic material.

Key historical evidence that DNA is the genetic material included the following:

- Chromosomes were first observed in the 1840s, and their association with heredity was observed in the early 1900s. Subsequent chemical analyses by biochemists showed that chromosomes are made up of DNA and protein, and that DNA itself is made of a five-carbon sugar (deoxyribose), phosphate, and four different nitrogenous bases.

- The nitrogenous bases are adenine (A) and guanine (G), double-ringed nitrogenous bases called *purines*; or thymine (T) and cytosine (C), single-ringed bases, called *pyrimidines*.

- In 1928, Frederick Griffith's efforts to develop a vaccine against a bacterial strain of pneumonia serendipitously showed that hereditary material was not inactivated by the death of the cell. When harmless bacteria were mixed with killed pneumonia-causing bacteria, the harmless bacteria were transformed into pneumonia-causing bacteria. The "transforming factor" was judged to be the genetic material.

- Oswald Avery, Colin MacLeod, and Maclyn McCarty (1944) conducted experiments to determine which part of the bacterial cell was the transforming factor. This team purified different components of killed disease-causing bacteria and mixed each component with harmless bacteria to see which component transformed the harmless variety. Their results indicated that DNA was the transforming material. Their conclusion was not widely accepted until subsequent experiments confirmed it.

- Martha Chase and Alfred Hershey (1952) performed definitive experiments that showed DNA—not protein—is the genetic material. Hershey and Chase infected bacteria with radioactively labeled T2 bacteriophages, viruses that infect bacteria. In a series of experiments the viral protein coat was labeled with radioactive sulfur (S35) and the nucleic acid inside was labeled with radioactive phosphorus (P32). The labeled viruses were then allowed to infect bacteria. Agitation of the infected bacterial cultures separated what was outside the cell from the actual bacterial cells. The sulfur labeled protein coat remained outside the bacterial cells and the phosphate-labeled viral nucleic acid was inside the bacterial cells. Only the viral material that enters the bacteria can be used to generate new viral particles, so it was DNA that acted on hereditary instructions.

- Erwin Chargaff (1949), a biochemist studying the DNA from many organisms, found that each kind of organism had a characteristic percentage of each of the four nitrogenous bases (A, G, T, and C), but that the amounts of each of the bases varied greatly from one kind of organism to another. This variability supported the possibility that DNA could contain enough information to be the genetic material. Chargaff also observed that within any species of organism, the percentage of bases A and T were equal to each other and the percentage of bases C and G were equal to each other. These two findings became known as *Chargaff's rules*.

- Rosalind Franklin, working with Maurice Wilkins (1952), generated X-ray crystallography images of DNA. These images indicated that DNA was a double helix with a uniform diameter consistent with pairing a purine nitrogenous base (A or G) with a pyrimidine base (T or C).

> **DID YOU KNOW?**
>
> Raymond Gosling was Rosalind Franklin's graduate student and is probably the person who actually took the famous X-ray diffraction photograph (known as Photo 51) that definitively showed that the structure of DNA was a double helix.

## DNA Structure

The structure of DNA was correctly modeled by James Watson and Francis Crick in 1953. Watson and Crick assembled a model of DNA that conformed to all of the experimental data that had been accumulating. Their model consisted of two DNA sugar-phosphate chains bound together by hydrogen bonds between complementary bases. The two strands then twist in a regular manner (once every 3.4 nm), hence, the term *double helix*.

Figure 14-1. DNA

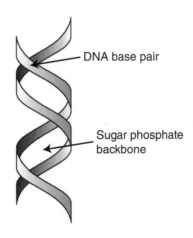

Nitrogenous bases, attached to each sugar, is one of the four nitrogenous bases composed of carbon and nitrogen rings. Chargaff's rule combined with Franklin's X-ray crystallography show that adenine pairs with thymine and guanine pairs with cytosine. This complementarity is the basis for correct replication and transcription.

**Figure 14-2. The Four Nitrogenous Bases**

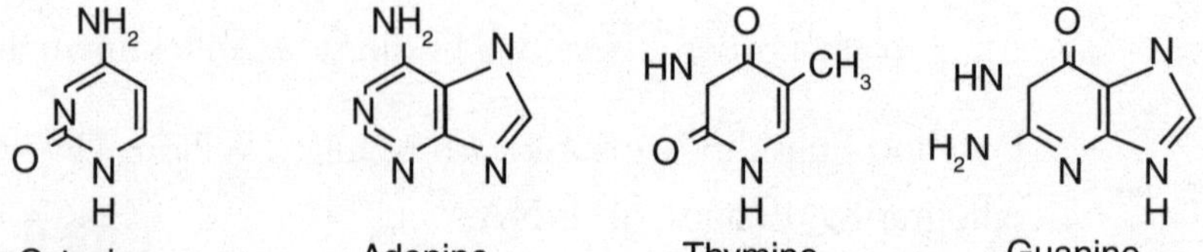

The complementarity between bases—that is, A always pairs with T, and C always pairs with G—contributes to DNA's ability to accurately replicate itself. The hydrogen bonding between bases contributes to DNA's stability.

## TEST TIP

For the AP Biology exam, you will need to understand the nature of complementarity and its role in ensuring faithful replication of DNA and transcription of mRNA.

**Figure 14-3. DNA Base Pairing**

Hydrogen bonds are indicated by dashed lines.

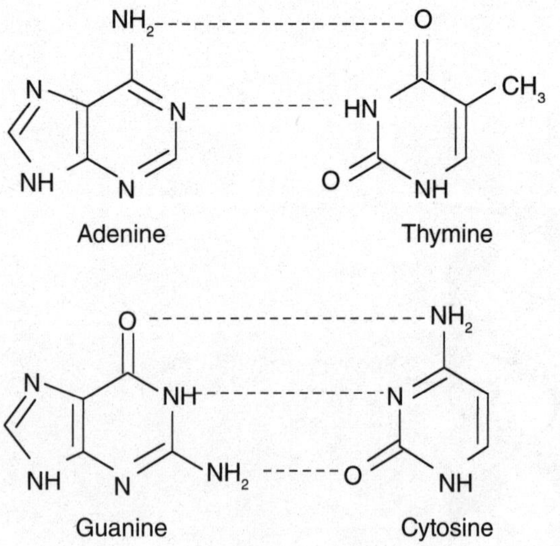

i. Adenine (purine) pairs with thymine (pyrimidine).
2 hydrogen bonds for base pairing.

ii. Guanine (purine) pairs with cytosine (pyrimidine).
3 hydrogen bonds for base pairing.

If one side of a DNA strand is known, its complement can be determined. For example:

5′GATTCGTAAGGC3′—one strand of DNA

3′CTAAGCATTCCG5′—its complementary strand

Because hydrogen bonds are relatively weak, Watson and Crick predicted that replication of DNA would occur when the strands of DNA are separated by enzymes, and each strand acts as a template for the synthesis of a complementary strand, forming two copies of the original double strand.

To better understand the structure of DNA, study Figures 14-4 and 14-5 and the accompanying explanations.

Figure 14-4. Adenine Nucleotide

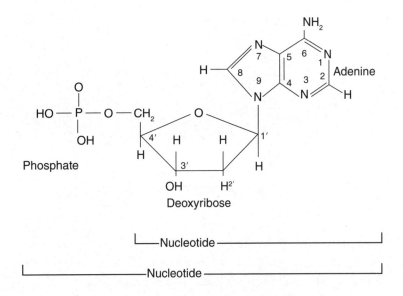

Deoxyadenosine-5′-monophosphate (dAMP)
(deoxyadenylic acid)

**Explanation of Figure 14-4:** The monomer units that make up DNA are nucleotides and each one is made up of a sugar, a phosphate group, and a nitrogenous base. Above is an adenine nucleotide (minus the phosphate group, it is a nucleoside). Other nucleotides would differ in the nitrogenous base present. Note the numbering of the carbon atoms of the sugar. New nucleotides can only be added to the 3′ end of a growing strand of DNA. This is important to know for replication and transcription.

Figure 14-5. DNA in Detail

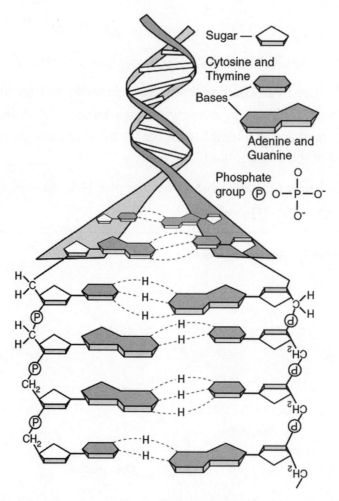

**Explanation of Figure 14-5:** One nucleotide is bonded to the next when the 5′ phosphate group of one nucleotide makes a bond with the 3′ carbon of the next nucleotide. This will result in a strand of DNA that will have the hydroxyl group of the 3′ carbon exposed at one end of the DNA strand and a phosphate group attached to a 5′ carbon exposed at the opposite end of the strand. The second strand will run antiparallel to the first strand when 2 hydrogen bonds form between complementary A and T and 3 hydrogen bonds form between complementary C and G.

# TEST TIP

You will not need to know any dates associated with the discovery of DNA—it is given for reference purposes only. You will, however, need to know the experimental contributions of Watson and Crick; Wilkins and Franklin; Avery, MacLeod, and McCarty; and Hershey and Chase. You may want to review these in greater detail.

# DNA Replication

The continuity of life is dependent on correct replication of genetic material. During the replication of DNA, each strand of DNA acts as a template for the synthesis of a new complementary strand. Experimentally confirmed by Matthew Meselson and Franklin Stahl (1958), this model is called *semiconservative replication*, versus conservative or dispersive replication. Meselson and Stahl grew *E. coli* on growth medium that contained heavy nitrogen, $^{15}N$, that the *E. coli* incorporated into their DNA as they grew on the medium. This "heavy" *E. coli* was then transferred to growth medium with only light nitrogen, $^{14}N$. Samples of *E. coli* were removed after one and then two rounds of replication. The DNA from these samples was extracted and centrifuged. After the first round of replication, the resultant banding pattern showed one band of medium density consistent with DNA that was made of half-heavy nitrogen and half-light nitrogen. This medium density DNA could have occurred if the method of DNA replication was dispersive or semiconservative, but not if replication was conservative.

The banding pattern that resulted from the *E. coli* that had undergone two rounds of replication showed two bands, one of medium density and one of light density. This pattern was consistent with a semiconservative model of DNA replication. Dispersive replication would have resulted in one band that was between medium and light.

The steps of DNA replication are initiation, elongation, and termination.

1. **Initiation** begins at an origin of replication, a specific sequence of nucleotides that can bind with proteins to create an initiation bubble. The bubble has a Y-shaped fork at each end called a replication fork. In bacteria, there is only one circular chromosome and only one origin. In eukaryotes, the chromosomes are linear and many origins can form within each chromosome. Enzymes called **helicases** unwind the helix at the origin(s) and break the hydrogen bonds that hold the strands together. Replication will proceed in both directions at each origin. **Topoisomerases** move ahead of each replication fork, relieving the strain caused by over winding.

2. **Elongation** occurs when **DNA polymerase** adds complementary nucleotides to the now-exposed strands of DNA. DNA polymerase can only add nucleotides to an exposed 3′ hydroxyl group, so a special enzyme, **primase**, places a short strand of RNA called a primer, on each exposed strand. The enzyme DNA polymerase can now attach complementary bases to the 3′ end of the primer. A DNA strand can only elongate in the 5′–3′ direction.

Because of the antiparallel nature of DNA, one strand of DNA can elongate continuously moving toward the replication fork. This strand is called the leading strand, and it only requires one primer sequence. However, the other strand, moving away from the replication fork, will elongate discontinuously requiring a new primer each time the fork opens. This strand is called the **lagging strand**. (See Figure 14-6.)

Another enzyme removes the RNA primers and replaces them with DNA nucleotides. Finally, DNA **ligase** joins the sugar-phosphate backbone into a continuous strand.

3. **Termination** occurs when one replication fork merges with another, comes to the end of the template strand, or does not have an RNA primer to add to. The primer at the 3′ end of the parent strand is removed, but cannot be replaced with DNA because there is no 3′ hydroxyl group to which to add a base. Thus, there is a bit of unreplicated end DNA with each round of replication. **Telomeres**—repetitious, noncoding DNA located at the ends of chromosomes—provide protection for coding DNA by providing noncoding "tails" that can go unreplicated without gene loss.

Complementarity ensures the accurate replication of DNA. However, nucleotide-matching mistakes occur. A particular DNA polymerase has the ability to proofread a growing strand, and remove and replace incorrectly matched bases. When these repair enzymes are faulty, DNA errors accumulate and can set the stage for disease.

Figure 14-6.  Replication Fork and Okazaki Fragments

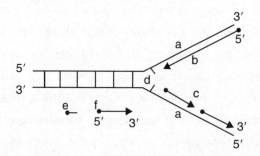

**Explanation of Figure 14-6:** A replication fork: (a) represents template DNA strands, (b) represents the leading strand, (c) the lagging strand, (d) is the replication fork. Each dot is an RNA primer, and f represents an **Okazaki fragment**—a short section of DNA preceded by a primer. Note that both strands are being synthesized in a 5′ to 3′ direction.

Table 14-1. Major Enzymes and Proteins in DNA Replication

| Enzyme | Substrate | Action |
| --- | --- | --- |
| DNA helicase | Double-stranded DNA | Opens up the DNA strand for replication |
| Single-stranded binding proteins | Single-stranded DNA | Binds single-stranded DNA and keeps replication fork open |
| DNA primase | Single-stranded DNA | Lays down an RNA primer on single-stranded DNA for DNA polymerase to hook up with |
| DNA polymerase | Single-stranded DNA | Adds the complementary base to the daughter strand using the parental template. Follows base-pairing rules: adenine with thymine, guanine with cytosine |
| DNA ligase | Single-stranded DNA | Links a 5' phosphate with a 3' hydroxyl on the lagging strand |

**TEST TIP**

You are expected to know the steps and the major enzymes involved in replication as well as the experimentation of Meselson and Stahl. In fact, rereading the experimentation of any of the scientists mentioned in this section will give you insight into science as a process.

## RNA Compared to DNA

Genes specify proteins through the two processes: **transcription** and **translation**. A second type of nucleic acid, **RNA**, carries out these processes. For most organisms, the flow of genetic information is from DNA to RNA to protein.

Table 14-2. Comparison of DNA to RNA

| Nucleic Acid | Structure | Bases Present/ Complementarity | Type/Function | Location |
|---|---|---|---|---|
| DNA | a. Double helix<br>b. Deoxyribose sugar | A-T; C-G | a. Directs DNA replication<br>b. Directs RNA synthesis<br>c. Directs protein production through RNA | Nucleus |
| RNA | a. Single strand<br>b. Ribose sugar | A-U*; C-G<br>*Note uracil is the complement to A in RNA. | a. mRNA carries transcript to cytoplasm for protein production.<br>b. rRNA is incorporated into ribosomes to carry out protein synthesis.<br>c. tRNA brings the correct amino acid to the ribosome. | a. Nucleus to cytoplasm<br>b. Cytoplasm<br>c. Cytoplasm |

# Transcription

The steps in transcription are similar to replication except a strand of mRNA carrying a DNA message is generated. **RNA polymerase** recognizes and attaches to a **promoter** sequence complexed with **transcription factors** on the template (coding) strand of DNA and unwinds and unzips this section. The promoter, which often includes a **TATA** box, determines what section of DNA will be the template.

Transcription occurs as RNA polymerase attaches complementary RNA nucleotides to the exposed template DNA until a termination sequence is reached in prokaryotes. It continues some distance past a termination sequence in eukaryotes.

The mRNA transcript in prokaryotes is functional immediately upon forming.

In eukaryotes, the initial transcript is altered in several ways. A 5′ cap and a poly-A tail are added that allow for attachment to a ribosome, facilitate transport of the mRNA out of the nucleus, and protect the transcript from degradation.

The eukaryotic mRNA contains coding regions called **exons**, and noncoding regions called **introns**. A splicosome, a large enzyme complex, will remove the introns and splice together the exons to make the final mRNA transcript. (See Figure 14-7.)

Figure 14-7. Splicing of mRNA to Remove Introns

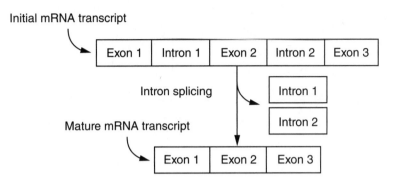

**Explanation of Figure 14-7:** The initial mRNA transcript contains introns that are excised and exons that are spliced together. With the cap and tail, the mRNA is ready for translation.

# Translation

**Translation** is the formation of a protein from a transcript. This process occurs in the cytoplasm and requires the collaboration of ribosomes (rRNA) and tRNA. The process also requires energy and specific enzymes.

Proteins are made of monomer units called amino acids. A given amino acid is specified by a three-nucleotide base sequence on the mRNA called a **codon**, which, in turn, is complementary to a three-nucleotide base sequence on the template DNA called a **triplet**. There are 20 different amino acids, each specified by one or more codons. For example:

**Template DNA** 3´ - TAC GGA AAT AAA TTC ATC-5´

**Spliced mRNA** 5´ - AUG CCU UUA UUU AAG UAG-3´

**Specified amino acid sequence** - Met Pro Leu Phe Lys Stop

The genetic code and its amino acid specificity is nearly universal to all organisms.

The codons on the mRNA are recognized by **tRNAs** carrying the specific amino acid required. The ribsosome, made up mostly of rRNA, is responsible for bringing codons on mRNA in contact with their specific charged tRNA for subsequent peptide bonding.

Figure 14-8. rRNA

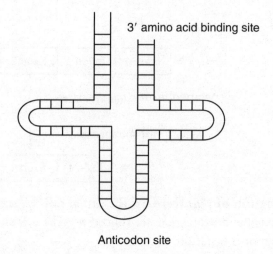

Anticodon site

**Explanation of Figure 14-8:** The anticodon site will be complementary to the codon on the mRNA within the ribosome. The codon will specify the amino acid, and the tRNA will insure the correct amino acid is presented.

**Ribosomes** are formed from two subunits, small and large, and these must come together with the mRNA to form a **translation unit**. Ribosomes can be free or bound. **Free ribosomes** are found in the cytoplasm. They complex with mRNA and proceed with translation. Protein formation occurs in the cytoplasm. As a protein forms, it carries a **signal peptide**. This is a sequence of amino acids that signal the ribosome the purpose of the protein—for staying in the cell or for export and/or for incorporation into a membrane. If an export signal peptide is recognized and the ribosome moves to the endoplasmic reticulum (ER), the ribosome is termed a **bound ribosome** and the forming protein moves directly from the ribosome into the lumen of the ER.

## Stages of Translation

There are three stages of translation: initiation, elongation, and termination.

1. **Initiation** begins when the mRNA leaves the nucleus and a ribosomal subunit attaches to it. The ribosome will scan the mRNA until it finds the start sequence, AUG. A tRNA with anticodon UAC, carrying a methionine amino acid, will attach to the codon. Then the larger subunit of the ribosome will attach and translation will begin.

2. **Elongation** of the polypeptide occurs as the mRNA moves through the ribosome. A tRNA complementary to the codon adjacent to the start codon will enter the ribosome and attach to it. A peptide bond will be formed between the two adjacent amino acids (see Figure 14-9). The ribosome will move the mRNA so the next codon translocates into the binding position. A charged tRNA will bring the specified amino acid, and a peptide bond will form. The first tRNA, minus its amino acid, will be released to the cytoplasm to be recharged. This process repeats itself forming a polypeptide which exits the ribosome.

3. **Termination** occurs when the mRNA presents a noncoding codon.

Figure 14-9.  Translation

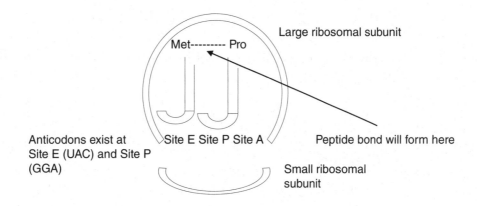

**Explanation of Figure 14-9:** After the tRNA carrying Met hydrogen bonds to the start codon, the large ribosomal subunit attaches to the small with the Met-bearing tRNA in the P site. The complex is ready to begin translation. The codon-specified tRNA will move into site A and a peptide bond will form between the amino acids in the P and A sites. The mRNA will translocate the mRNA so the tRNA in the P site moves into the E site. The tRNA in the A site translocates to the P site and a charged tRNA will move into the A site and peptide bonding will occur. This will continue until a termination codon is reached. In the example, the Met bearing tRNA will be released to be recharged; site A is awaiting a charged tRNA.

More than one ribosome can read a transcript at a time. As the polypeptide exits the ribosome, it begins to naturally fold based on the environment and the primary structure of the polypeptide. Final folding and refinements to the protein may be necessary.

Protein production is predicated on correct DNA. A mutation is a change in DNA and it has the potential to change a protein. The change may be tolerated if it does not affect the protein function; however, it is more likely that a mutation will change or stop the protein function. **Point mutations** are single base pair DNA changes such as:

- **Substitution mutations** occur when a coding base is changed that results in a change in the amino acid sequence of the protein:

    Original template        DNA 3′ - TAC GGA AA**T** AAA TTC ATC-5′

    Original mRNA            5′ - AUG CCU UU**A** UUU AAG UAG-3′

    Specified amino acid sequence Met Pro **Leu** Phe Lys Stop

    Subtitution mutation:    DNA 3′ - TAC GGA AA**A** AAA TTC ATC-5′

    Altered mRNA             5′ - AUG CCU UU**U** UUU AAG UAG-3′

    Altered amino acid sequence – Met Pro **Phe** Phe Lys Stop

    The faulty shape of sickle cell hemoglobin is due to a base-pair substitution mutation.

- The DNA is read as a triplet sequence so an insertion or deletion of a base would *shift* the reading. This point mutation is known as a **frameshift mutation.**

    Original template DNA 3′- TAC G**G**A AAT AAA TTC ATC-5′

    Original mRNA 5′ - AUG CCU UUA UUU AAG UAG-3′

    Amino acid sequence Met Pro Leu Phe Lys Stop

    Deletion of G DNA 3′- TAC GA AAT AAA TTC ATC-5′

    Resultant mRNA 5′ - AUG CUU UAU UUA AGU AG-3′

    Altered amino acid sequence Met Leu Tyr Leu Ser … changed

- An addition would *shift* the reading frame in a similar manner to deletion. These types of mutations are almost always extremely negative. Any mutation that changes an amino acid is a **missense mutation**, and, if the amino acid substitution is a stop codon, it is said to be a **nonsense mutation**.

Table 14-3. The Genetic Code

| | | Second Position of Codon | | | | | |
|---|---|---|---|---|---|---|---|
| | | U | C | A | G | | |
| **F I R S T   P O S I T I O N** | U | UUU Phe<br>UUC Phe<br>UUA Leu<br>UUG Leu | UCU Ser<br>UCC Ser<br>UCA Ser<br>UCG Ser | UAU Tyr<br>UAC Tyr<br>UAA Ter [stop]<br>UAG Ter [stop] | UGU Cys<br>UGC Cys<br>UGA Ter [stop]<br>UGG Trp | U<br>C<br>A<br>G | **T H I R D   P O S I T I O N** |
| | C | CUU Leu<br>CUC Leu<br>CUA Leu<br>CUG Leu | CCU Pro<br>CCC Pro<br>CCA Pro<br>CCG Pro | CAU His<br>CAC His<br>CAA Gln<br>CAG Gln | CGU Arg<br>CGC Arg<br>CGA Arg<br>CGG Arg | U<br>C<br>A<br>G | |
| | A | AUU Ile<br>AUC Ile<br>AUA Ile<br>AUG Met | ACU Uhr<br>ACC Thr<br>ACA Thr<br>ACG Thr | AAU Asn<br>AAC Asn<br>AAA Lys<br>AAG Lys | AGU Ser<br>AGC Ser<br>AGA Arg<br>AGG Arg | U<br>C<br>A<br>G | |
| | G | GUU Val<br>GUC Val<br>GUA Val<br>GUG Val | GCU Ala<br>GCC Ala<br>GCA Ala<br>GCG Ala | GAU Asp<br>GAC Asp<br>GAA Glu<br>GAG Glu | GGU Gly<br>GGC Gly<br>GGA Gly<br>GGG Gly | U<br>C<br>A<br>G | |

**Explanation of Table 14-3:** This table is based on mRNA codons. Notice that many of the amino acids are coded for by more than one codon. The genetic code is said to be redundant.

# TEST TIP

You are not expected to memorize the codon chart. A codon chart will be provided on the AP test if you are asked to translate a particular segment of DNA or mRNA.

## Questions to Consider

1. What would happen to the structure of DNA if Chargaff's rules were violated. For example, what would happen if A bonded with G?

2. Look carefully at the structure of DNA. What kinds of bases are you more likely to find an abundance of at an origin of replication?

3. How might a point mutation change the folding of a protein?

## Answers:

1. DNA is a double helix with a constant diameter. This is only possible if a purine (A or G) bonds with a pyrimidine (T or C). Purine is a double-ring structure, so it is larger than the single-ring structure of pyrimidines. If an A bonded with a G, the diameter of the helix would bulge because both A and G are larger structures.

2. There are three hydrogen bonds between C and G and two hydrogen bonds between A and T. It is likely that an origin of replication would be an area with an abundance of A–T bases because these would have relatively fewer hydrogen bonds that would need breaking. An interesting fact is that thermophilic prokaryotes have DNA that is rich in C–G bonds.

3. The primary sequence of a protein is the correct order of amino acids and it is this sequence that determines the initial pattern of folding for the protein. For example, hydrophobic amino acids will fold inwards and hydrophilic amino acids will protrude into the watery surroundings. If a point mutation changes an amino acid and the change is from hydrophilic to hydrophobic, for example, the folding of the protein in the vicinity of the amino acid change may dramatically affect the protein-folding pattern.

# Cell Cycle, Mitosis, and Meiosis

Chapter 15

## Overview

The continuity of life is dependent on reproduction. Since organisms are made up of cells and cells come from cells, the continuity of life actually begins with cell division.

Cell division is part of the **cell cycle**, a series of regulated stages in the life cycle of a cell. The cell cycle starts with a new cell and proceeds through to formation of daughter cells. There are two types of cell division among eukaryotes. — nucleus

**Mitosis** is a cell division that results in two genetically identical daughter cells and occurs in the somatic or non-**gamete** producing cells of an organism.

**Meiosis** is a special two-stage cell division restricted to gamete producing cells that results in four daughter cells, each of which has half the number of chromosomes characteristic of the organism.

## The Cell Cycle

For many organisms, some form of cell division is the way the organism reproduces. For most multicellular, eukaryotic organisms, cell division makes more cells to repair injury, replace damaged cells, or to allow organisms to grow.

DNA is contained within chromosomes. In eukaryotes, chromosomes are long strands of DNA wound around proteins called histones and organized into nucleosomes. The collection of DNA and proteins within the nucleus of the cell is chromatin.

Each eukaryotic organism has a characteristic number of chromosomes within its nucleus. For example, humans have 46 chromosomes and goldfish have 94. These chromosomes exist as 2 *sets* of chromosomes, one set from the female parent and one set from the male parent. The sets exist as matched pairs, that is, instructions for the same hereditary feature are found on a pair, but with each parent potentially contributing slightly different instructions for the trait (remember alleles). Each matched pair of chromosomes is called a **homologous pair**. The term **diploid**, 2n, is used for cells that have both sets of chromosomes. The term **haploid**, n, describes cells that have one set of chromosomes.

Prior to either mitosis or meiosis, each chromosome is duplicated and then highly condensed. This facilitates movement of the chromosomes during nuclear division.

Figure 15-1.  Replicated Chromosome Prior to Mitosis or Meiosis; Held Together at the Centromere

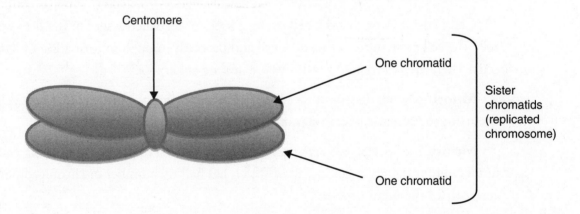

The cell cycle is a sequence of events in the life of a cell that includes **interphase**, **mitosis**, and **cytokinesis**. Not all cells will undergo the complete cycle. If a go-ahead signal is not given to a cell, it will exit the cell cycle and enter $G_0$. Many somatic cells remain in $G_0$ for their entire life cycle.

**Interphase** is the longest part of the life cycle of a cell and the time during which it is carrying out its normal functions. If a cell is signaled to undergo cell division, additional events will occur during interphase. Figure 15-2 shows the phases of the cell cycle.

Figure 15-2. The Cell Cycle

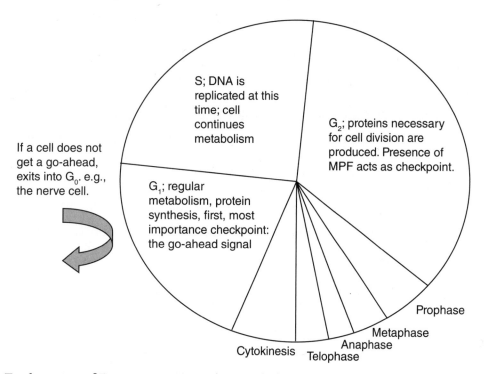

**Explanation of Figure 15-2:** Interphase includes $G_1$, S, and $G_2$. Prophase through telophase represents mitosis; cytokinesis is the final cell division that will result in two daughter cells and completes the cell cycle.

**Mitosis** is nuclear division that separates and distributes the replicated chromosomes into two daughter cells. The four phases of mitosis are **prophase, metaphase, anaphase,** and **telophase.** Use these explanations as you review Figure 15-3 on the next page.

- In **prophase (1, 2)**, the replicated chromosomes condense and coil, becoming visible as individual strands under a light microscope. The nuclear membrane breaks down; microtubules in the cytoplasm are breaking down and reassembling as spindle fibers. In animal cells, 2 pairs of centrioles are moved by microtubules to opposite sides of the nucleus.

- In **metaphase (3)**, the microtubules interact with the chromosomes moving the chromosomes into alignment at the equator of the cell. The microtubules are attached to the kinetochores of the centromere.

- During **anaphase (4)**, the shortest phase of mitosis, the sister chromatids begin to separate as the cohesion proteins that held them together at the centromere are broken. Microtubule shortening pulls the now-separated chromosomes apart. The cell elongates. The phase ends when each end of the cell has a complete set of chromosomes.

- **Telophase (5)** is the last stage of mitosis. The nuclear membrane reforms, nucleoli reform, and the chromosomes decondense to chromatin.

**Cytokinesis (6, 7)**, the division of the cytoplasm and its organelles, starts during telophase. In animal cells, cytokinesis occurs when a ring of actin protein pinches the cytoplasm in two. In plant cells (not shown in Figure 15-3), cytokinesis begins with the production of a cell plate across the center of the cell. This cell plate forms from the center out.

Figure 15-3.  The Phases of Mitosis

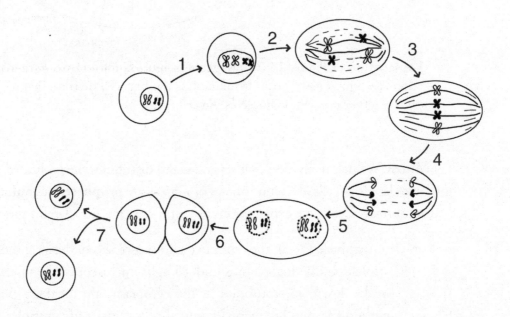

Regulation of the cell cycle occurs via signals that initiate and/or manage the progress of the cell cycle. The cell cycle has checkpoints that prevent faulty cells from undergoing mitosis. There is a checkpoint during $G_1$, the restriction point, which determines whether a cell will proceed through mitosis. Once past this checkpoint, other checkpoints monitor the correct progress of the cycle. One internal control mechanism involves the changing concentrations of cyclins within a cell. Cyclins are proteins that activate cyclin-dependent kinases to form MPF (mitosis promoting factor), which triggers continuation of the cycle past $G_2$. A checkpoint in anaphase ensures that all kinetochores are attached to a microtubule.

External signals, such as growth factors released after an injury, can set the cell cycle in motion by setting off a **signal-transduction pathway**. Most cells' division activity is inhibited by the presence of other cells (density inhibition) or the absence of a substrate.

Any changes in the response to internal or external cell cycle cues can cause cells to become cancerous by allowing uncontrolled cell division or division of faulty cells.

# Meiosis

In a sexual life cycle, meiosis is a two-cycle cell division that reduces the chromosome number from diploid (2n) to haploid (n) by separating homologous chromosome pairs into cells that produce gametes.

Meiosis produces four daughter cells, each with a set of chromosomes, that is, one member of each pair of homologous chromosomes (one from the mother, one from the father, etc.). The haploid gametes are restored to diploid through fertilization. As in mitosis, the cells that are to undergo meiosis will replicate their chromosomes during interphase. This will be followed by two rounds of cell division: Meiosis I (steps 1-5 in Figure 15-5), the reduction division, and Meiosis II (steps 6-10 in Figure 15-5).

During prophase I (Figure 15-5, steps 2 and 3), replicated chromosomes condense and homologous pairs synapse. During synapsis, the homologous pairs become physically connected by proteins and exchange genes by crossing over. In Figure 15-4 the homologous pairs, A and a, are crossing over (I) and genetic exchange has occurred in (II). This event does not occur in prophase of mitosis. In addition, during prophase I, there is spindle formation and nuclear membrane breakdown in a manner similar to mitosis.

### Figure 15-4. Crossing Over During Prophase I

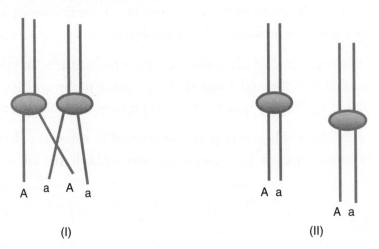

**Explanation of Figure 15-4:** Homologous pair of chromosomes synapsing (I). The synapsing homologous pair is called a tetrad. Note that alleles A and a from nonsister chromatids have crossed over (II). The crossing-over event greatly increases variation in gametes.

During metaphase I (Figure 15-5, step 3), *homologous pairs* line up along the equator. When the pairs line up on the equator, they do so randomly. Microtubules become attached to each member of a homologous pair at a kinetochore.

In anaphase I (Figure 15-5, step 4), the homologous pairs (tetrads) are separated. Centromeres do not split as in mitosis. Since the homologous pairs lined up randomly along the equator during metaphase, the homologous pairs are independently assorted during anaphase I.

Meiosis I is the reduction division because of the separation of the homologous pairs resulting in haploid daughter cells.

Telophase I and cytokinesis I (Figure 15-5, step 5) occurs, and the results are two daughter cells that are haploid.

Each of these two haploid daughter cells now proceed through Meiosis II (Figure 15-5, steps 6-10). Much like in mitosis, a spindle forms and the chromosomes line up singly and randomly along the equator with microtubules attached to kinetochores on each sister chromatid. After the centromere proteins split, sister chromatids are separated. The result is four haploid, unique cells from one initial cell.

Figure 15-5.  Meiosis

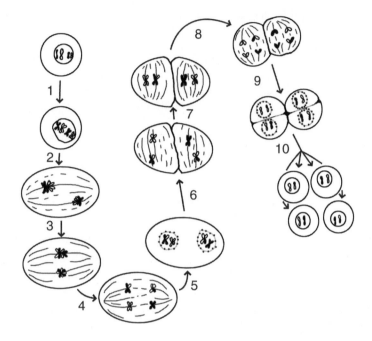

Several steps of meiosis provide opportunities for genetic variation and/or changes to chromosome structure. The processes of **crossing over** and **random assortment** naturally make each gamete unique. These are two critical ways that sexual reproduction increases variation in offspring. **Random fertilization**—any given sperm with its possible variations can combine with any given egg with its possible variations—is another.

Other changes can occur during meiosis that potentially introduce a mutation or change in the DNA of an organism. For example, **duplications** or **deletions** can occur when a chromosome breaks during meiosis—usually during a crossing-over event. The broken piece of chromosome may not rejoin its chromatid, resulting in a deletion on that arm of the chromosome. The broken piece may become incorporated into its homologous chromosome resulting in a duplication on that arm. The broken piece may also invert and reattach to its chromatid as an inversion, or it may become translocated to a nonhomologous chromosome. These events are almost always harmful. See Figure 15-6 for an illustration of each of these events.

Figure 15-6. Mutations Involving Changes in Chromosome Structure

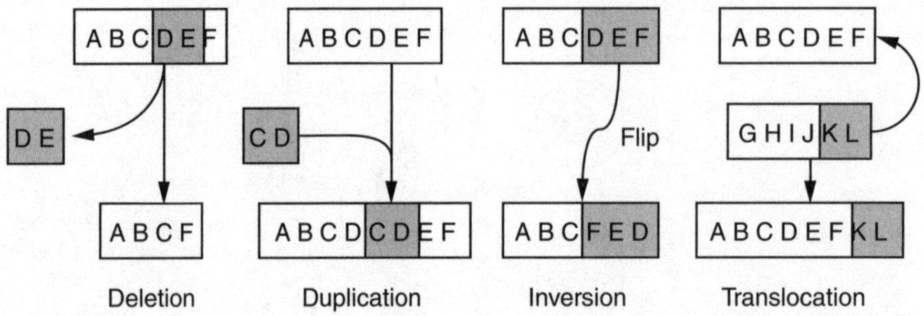

|  |  |  |  |
|---|---|---|---|
| ABCDEF → ABCF | ABCDEF → ABCDCDEF | ABCDEF (Flip) → ABCFED | ABCDEF + GHIJKL → ABCDEFKL |
| Deletion | Duplication | Inversion | Translocation |

Changes in an organism's chromosome number can occur during meiosis. Failure of one or more homologous pairs or sister chromatids to separate is called nondisjunction. If a nondisjunction occurs during anaphase I and homologous pairs of chromosomes fail to separate, or if sister chromatids fail to separate during anaphase II, gametes will result that have too many and too few chromosomes. Trisomy 21, Down syndrome, is an example of a condition that usually results from nondisjunction of chromosome 21. An egg (usually) or sperm with an extra chromosome 21 due to nondisjunction is joined by a normal egg or sperm and will result in a fertilized egg with three chromosome 21.

> **DID YOU KNOW?**
>
> Another form of trisomy 21 (Down syndrome) occurs through a translocation event during meiosis. A piece of chromosome 21 is translocated to a nonhomologous chromosome resulting in a gamete with an extra bit of chromosome 21 and a normal number 21 chromosome. If that gamete is fertilized by a normal gamete, the karyotype will have the normal 46 chromosomes but the individual will have Down syndrome because of the extra bit of translocated 21.

When meiosis fails to occur after chromosome replication, gametes may result that are still 2n. If these gametes are joined by another diploid gamete, a viable 4n zygote may result that has the potential of reproducing with like 4n individuals. This is a form of **polyploidism** called **autopolyploidism**. Many new plant species have arisen as a result of autopolyploidism.

## TEST TIP

If you are asked how many chromosomes are in a cell, count the centromeres *not* the chromatids.

# The Chromosomal Basis of Inheritance

DNA and its association with genes and chromosomes was not understood well until later into the twentieth century. However, in 1860, Gregor Mendel worked out the basic principles of inheritance through his experimentation with garden peas. Mendel developed the principle of dominance, the law of segregation, and the law of independent assortment. Mendel also used laws of probability to predict the inheritance patterns that could be expected from a particular genetic cross. It is now known that many inheritance patterns are not as straightforward as predicted by Mendel.

## Mendel's Experiment

Mendel chose to experiment with garden peas that had characteristics that showed clear contrasting traits. This was a fortunate choice because peas were cheap to work with and generated a lot of data.

Mendel's initial experiments were carefully conducted crosses between plants with one characteristic that had two contrasting forms. For example, one cross was between pure-breeding tall plants and pure-breeding short plants. A cross that considers only one character is called a monohybrid cross.

In each monohybrid cross, the first generation of plants, now termed the $F_1$ generation, showed only one of the original two traits. In our example, all the $F_1$ plants were tall. Mendel carefully mated the $F_1$ plants to generate a second generation, the $F_2$ generation. In each cross, the trait that had "disappeared" in the $F_1$ reappeared in 25% of the $F_2$ plants. That is, 75% of the $F_2$ plants were tall and 25% were short.

These observations led Mendel to propose that all characteristics are governed by two traits, one trait from the mother and one from the father. If one trait "disappeared," it was said to be recessive and the trait that masked its presence was termed dominant. Mendel proposed that each characteristic (gene) pair was separated in the process of making eggs or sperm. This is Mendel's law of segregation. We now term the characteristics being considered (height) a gene and the different traits that can be expressed (tall or short) alleles.

The Punnett square is a common device used to work out simple genetic crosses. Note that one parent's genetic contributions go across the top, the other down the side. The boxes represent all the possible offspring allele combinations for the given gene. If an allele is dominant, the heterozygote (Tt or tT) form will appear the same as in the pure-breeding parent that displays the dominant trait.

Figure 15-7. Punnett Square for the Cross of Heterozygous Tall Plant

|   | T  | t  |
|---|----|----|
| T | TT | Tt |
| t | tT | tt |

Mendel went on to do careful crosses involving two characteristics such as height and pod color. These kinds of crosses are called dihybrid crosses. The following is an example of one of Mendel's dihybrid crosses:

Starting generation (**parental**)

Pure breeding Tall(TT)Green (GG) ×φ Pure breeding Short (tt) Yellow(gg)

$F_1$:   All Tall (Tt) Green (Gg)

$F_1 \times F_1$ in a dihybrid gives a characteristic ratio in the $F_2$:

$\frac{9}{16}$ Tall Green (T_G_): $\frac{3}{16}$ Tall Yellow (T_gg): $\frac{3}{16}$ Short Green (ttG_); $\frac{1}{16}$ Short Yellow (ttgg)

From these observations, Mendel developed his law of independent assortment. This means that characteristics (genes) separate independently of each other. In our example, the gene for height was inherited independently of the gene for color.

## Using the Laws of Probability in Genetics

Mendel's inheritance patterns correspond to rules of probability. Two of the most important rules are the *multiplicative rule* and the *additive rule*. The multiplicative rule says that the chance of two independent events occurring simultaneously is the product of the individual chances. In our example, TtGg × TtGg are crossed. What is the probability of producing tall green peas from this cross?

$$T\_ = \text{tall}; Tt \times Tt = \tfrac{3}{4} T\_$$

$$G\_ = \text{green}; Gg \times Gg = \tfrac{3}{4} G\_$$

$$T\_G\_ = \frac{3}{4} \times \frac{3}{4} = \frac{9}{16}$$

The additive rule states that if an event can occur in more than one way, its probability of occurrence is the sum of the different ways the event can occur. For example, in a cross of Tt × Tt, a heterozygote can occur as tT or Tt, depending on which parent contributes which particular allele. So, by the additive rule, in this case, Tt = ¼ tT + ¼ Tt = ½ chance of heterozygote.

The multiplicative rule calculates the probability of independent events occurring when the genes are linked. For example, in a cross of AaBbCc × AabbCC, what are the chances of the AabbCC progeny? The answer is Aa × Aa = ½ Aa; Bb × bb = ½ bb; Cc × CC = ½ CC. Therefore, for the progeny AabbCC, there is a ½ × ½ × ½ = ⅛ probability.

## Non-Mendelian Genetics

Non-Mendelian genetics refers to instances when genetic characteristics do not follow inheritance patterns of Mendel's pea experiment. Genes govern the production of proteins. The expression of the proteins governs the *possible* phenotype of the individual. The actual phenotype expressed is affected by contributions of the environment and/or other gene expressions as well as by the physiology of the individual. Some examples of non-Mendelian genetics include incomplete dominance and codominance.

Incomplete dominance requires that some inheritance is controlled by more than two alternative alleles. For example, most human blood types are under the control of three alleles, the allele for A blood type, the allele for B blood type, and the allele for O blood type. A and B are codominant and will be equally expressed and O is recessive to both A and B.

Not all inheritance is expressed as simply dominant or recessive. Codominance occurs when the two alleles present in an individual contribute to the ultimate phenotype. An example is the human blood type AB, where the individual has an allele for A and an allele for B and each is expressed as a particular glycolipid on the surface of the blood cells. Sometimes an allele contributes nothing to the phenotype, so only half the protein product of the gene is produced in the individual. For example, red × white flowers may yield pink flowers because one allele codes for red pigment and one allele produces no pigment. The resulting phenotype is pink because the individual is producing half the pigment that a pure red plant would produce.

Most phenotypes result from the action of multiple genes (polygenic) that act in an additive manner. Skin color can range from very light to very dark dependent on the number of light vs. dark pigment genes inherited.

Polygenic inheritance occurs when many characteristics are inherited as a continuous inheritance in which many alleles contribute to a range of phenotypes. Some examples in humans include eye color, skin color, and cancer.

Genomic imprinting is a physiological process that methylates (adds methyl groups) certain genes. This occurs during gamete formation. The imprinting differs based on whether the imprinting occurs during egg production or sperm production. So, the phenotype expressed in the offspring is dependent on which parent contributes the expressed allele and which contributes the imprinted allele. This is not a sex-linked trait as the imprinting occurs on somatic chromosomes.

Some subcellular organelles, such as chloroplasts in plants and mitochondria in both plants and animals, have DNA that is independent of the nuclear DNA. This organelle DNA is circular and carries genes that produce proteins used within the organelle.

- The DNA in organelles is circular vs. linear. This is evidence for endosymbiosis.

- The organelles and the DNA in them are replicated and distributed to new daughter cells when the cells undergo mitosis and meiosis.

- Organelle DNA can mutate and pass on accumulated abnormalities.

Mitochondrial DNA contains genes that produce proteins associated with cellular respiration. For most sexually reproducing organisms, mitochondrial DNA is inherited from the mother because the sperm is very small and contributes mainly genetic material. Cytoplasmic contents of the fertilized egg come from maternal gametogenesis. This means that any disorders associated with mitochondrial DNA are inherited from the mother. This is not sex-linkage as all of the offspring, male or female, inherit maternal mitochondria. Mitochondrial disorders are generally felt most in the systems that required the most energy, such as the nervous system. Over the lifetime of an organism, mitochondrial DNA can accumulate mutations. It is thought these may contribute to disorders associated with aging.

> **DIDYOUKNOW?**
>
> Reginald Punnett developed the Punnett square after Mendel's work surfaced in the early 1900s. Punnett could not understand why a dominant gene would not become fixed within a population. He explained his problem to his mathematician colleague, G.H. Hardy, who went on to develop the Hardy-Weinberg Principle, which helps to explain gene frequencies within a population.

Some human genetic disorders exhibit Mendelian inheritance. For example, cystic fibrosis and sickle cell anemia are inherited as recessive disorders in humans. Individuals who are heterozygous and pass on the recessive gene are said to be carriers. Very few human genetic disorders are dominant with only one affected parent being necessary to pass on the disorder. An example is Huntington's disease.

## Chromosome Theory of Inheritance

The link between Mendel's inheritance patterns, the pattern of chromosome movement during the cell cycle, and the knowledge that DNA is the genetic material and is located on chromosomes constitutes the chromosome theory of inheritance.

**Linked Genes**—T. H. Morgan's experiments with fruit flies provided evidence to support the chromosome theory of inheritance when he discovered linkage. That is, each chromosome contains more than one gene and, if genes are present on the same chromosome, they are not going to sort independently, but travel as a group. Genes that are linked on a chromosome undergo crossing over (recombination) during prophase I. The recombinant events are proportional to the distance between the genes. Genes that are closer on a chromosome are inherited together more often that genes than are farther apart. This data can be used to map a chromosome.

**Sex-Linked Genes**—Sex in humans and in many organisms is determined by a pair of nonhomologous chromosomes. Inheritance of two X chromosomes yields a female; inheritance of an X and Y chromosome yields a male. The Y chromosome is small and carries a regulatory gene that sets in motion production of a male. The X chromosomes are larger and carry more genetic material. This is the basis for sex-linked disorders such as colorblindness.

## Ethical Issues of Genetics

Many ethical issues have arisen with increased understanding of genetics and the ability to correlate genes to disorders or to predisposition for disorders. Some of the ethical issues include inappropriate application of genetic testing (for example, selecting the sex of offspring), inappropriate cloning of organisms, and setting gene patent laws.

The Genetic Information Nondiscrimination Act of 2008 initiated a federal law that prohibits discrimination in healthcare coverage based on genetic information. This law, however, does not prohibit insurers from using genetic information to set rates.

Ethical questions arise when doctors treat patients for a disorder with a known genetic predisposition. What are the doctors' obligations or prohibitions on informing relatives who might share the same genetic predispositions?

Some genetic tests provide reliable information to enable informed decision making. However, many genetic tests cannot give reliable information about the onset or severity of a genetic condition, thus there are limitations to genetic tests that have the potential to cause undue stress.

## TEST TIP

You will need to be able to work monohybrid and dihybrid genetics problems and use the product rule and additive rule to predict the outcomes of more involved crosses.

## Questions to Consider

1. In humans, pseudohypertrophic muscular dystrophy is a condition in which the muscles gradually waste away, ending in death in the early teens. It is dependent upon a sex-linked recessive gene. This type of muscular dystrophy occurs only in boys and has never been reported in girls. Why is it not to be expected in girls?

2. A man with blood type O is also Rh negative. These are both completely recessive inherited traits. He marries a woman with blood type A who is Rh positive, both dominant characteristics. These two genes, blood type and Rh status, are not linked genes. This couple has four children, all with blood type A but two of the children are Rh negative. What is the mother's probable genotype?

## Answers:

1. Affected males usually die before they are able to pass on their gene to a daughter. A girl can be a carrier, but not be affected because she would need to receive the dystrophy allele from both parents to be affected.

2. We know the man's genotype is homozygous for both blood type and Rh factor because both were recessive. We know that blood type A is completely dominant to O, and Rh positive is completely dominant to Rh negative. Since all the children are type A blood, it is likely the mother is homozygous for A, *but* because some of the children are Rh negative, the mother must be heterozygous for Rh factor.

Father: ii rr X $I^AI^A$ Rr Mother

Children: $I^A$iRr or $I^A$irr

Note that rr × Rr gives 50% Rr:50%rr

**Time for a quiz**
- Review strategies in Chapter 2
- Take Quiz 5 at the REA Study Center

*(www.rea.com/studycenter)*

# Chapter 16

# Regulation of Gene Expression and Genetic Variation

## Overview

Organisms control expression of their genes in response to changes in the external and internal environments that they encounter. Single-celled organisms (prokaryotes), such as bacteria, can conserve energy and resources—giving them an evolutionary advantage—by producing only those proteins necessary to respond to their existing environment. Multicellular organisms (eukaryotes) must be able to regulate genes to adjust to an immediate environment, but they must also be able to regulate genes that coordinate the development of an overall body plan.

**Gene regulation** in bacteria involves control over the transcription of operons, and in eukaryotes it involves multiple levels before, during, and after transcription and translation.

Regulatory proteins can provide both positive and negative control mechanisms for gene expression. This gene expression can be inhibited by binding to DNA and blocking transcription. This is called negative control. Then, gene expression can be stimulated by binding to DNA and then stimulating transcription, called positive control.

# Gene Regulation in Prokaryotes

Bacteria have a single double-stranded, circular chromosome not enclosed within a membrane. Separate, small rings of DNA often found in bacteria are called **plasmids**. These do not hold genes critical to bacterial survival, but they are replicated with each cell replication. The genes found on plasmid may be helpful and are expressed by the bacteria. Plasmids are an important source of genetic variation in bacteria and are important as vectors in biotechnology.

Bacterial genes are clustered into functional units called **operons**. Each operon has a regulator gene, a promoter, an operator, and the genes necessary to produce the proteins of a particular metabolic pathway. The genes are not interrupted by noncoding DNA. There are two kinds of operons—**inducible** and **repressible**.

Bacterial regulation usually takes place at the transcription level and is controlled by a repressor protein coded for by a regulator gene located upstream from the rest of the operon. An active repressor protein binds to the operator to shut down transcription by blocking RNA polymerase from attaching to the promoter. A **co-repressor** is a molecule that binds to a repressor protein and either activates it or makes it inactive.

Table 16-1. Comparison of Inducible vs. Repressible Operons

| Operon Type | General Action | Co-repressor | Result | Example |
| --- | --- | --- | --- | --- |
| Inducible | *Catabolic:* Breaks down molecules, often for energy | Makes repressor inactive | Transcription proceeds | *Lac* operon breaks down lactose for energy. |
| Repressible | *Anabolic:* Synthesizes necessary molecules not present in surroundings | Activates repressor | Transcription halted | *Tryp* operon builds tryptophan in its absence. |

Both inducible and repressible operons operate by **negative feedback** inhibition. That is, if the repressor protein is active, transcription of the metabolic pathway proteins is stopped. Many metabolic pathways in both prokaryotes and eukaryotes are regulated by negative feedback inhibition.

Some genes are regulated by **positive feedback**. That is, an activator stimulates increased gene transcription. An example of positive feedback occurs in bacteria when cAMP, a substance that builds up as ATP is consumed, binds with an activator protein that attaches to an operon, increasing transcription efficiency.

In prokaryotes, because there is no nuclear membrane, **ribosomes** attach and translate the RNA transcript as it is being synthesized.

> **TEST TIP**
>
> Before you go into the AP Biology test, know at least one example of negative and positive feedback. Feedback control of metabolic pathways is an important concept.

## Gene Regulation in Eukaryotes

In eukaryotes, gene control is more complicated than prokaryotic control. One reason is that there are more opportunities for control than in prokaryotes. Most gene expression, prokaryote or eukaryote, is controlled by controlling transcription.

Eukaryotes have linear chromosomes confined within a nucleus. All cells in a multicellular organism—except for gametes and red blood cells—have all of the organism's genes. Some housekeeping genes within a cell are expressed in all cells at all times; some genes are present and active only as a cell undergoes differentiation; and some genes are expressed as conditions change, such as in response to a stimulus.

Initial **gene regulation** occurs by controlling access to DNA such that transcription cannot occur. DNA is wound around proteins called **histones** to form **nucleosomes**. Nucleosomes can be further wound and folded back on each other. Packing DNA enables a large volume of material to be fit into a nucleus. The ultimate condensation of a chromosome occurs during metaphase in either mitosis or meiosis. The more condensed DNA is, the less accessible it is for transcription.

Once DNA is wound around the histone, chemical changes occur that maintain the nucleosome structure. Methylation, addition of methyl groups to histone tails, increases DNA concentration and blocks access of transcription factors to DNA. Methylation of DNA itself, especially of promoter regions of DNA, can also limit transcription. Conversely, acetylation of histones loosens their hold on DNA

> **DID YOU KNOW?**
>
> In female mammals, one of the two inherited X chromosomes is inactivated. It is highly condensed and methylated and remains inactive inside the nuclear envelope. This chromosome can be seen under a microscope and is referred to as a Barr body. It ensures that both males and females each operate with only one copy of the genes found on an X chromosome.

and makes it more accessible for transcription. Modifications to chromatin are referred to as **epigenetics** and do not involve changing the DNA.

In eukaryotes, a segment of DNA that codes for a protein or polypeptide typically consists of introns and exons, a transcription start site, a general promoter that is near the DNA start site and includes the TATA box and some specific control elements, and a more distant control element called an enhancer. Transcription factors bind to the general promoter and to each other to form a complex that RNA polymerase II can recognize and bind to begin transcribing DNA. This results in a relatively slow rate of transcription. The rate is increased when the distant enhancer interacts with specific transcription factors. When the enhancer binds with transcription factors, the DNA bends, forming a loop. The activated enhancer region binds with the proteins bound to DNA in the promoter region, making a complex that is readily recognized by RNA polymerase. And, the rate of transcription rises dramatically.

**Transcription** can be halted or prevented by a repressor protein that binds to a silencer region. This is an area of DNA near an **enhancer region**. When the silencer region is bound by a repressor, transcription factors cannot bind to the enhancer region. The particular combination of specific transcription factors in the enhancer region and promoter site controls what gene will be transcribed. The same combination of control elements is probably responsible for coordinating the expression of genes that must be transcribed and expressed at the same time.

**Post-transcriptional control** occurs in several ways. Alternative RNA splicing can generate different mRNAs from the same initial transcript. mRNA has a relatively short lifespan before degradation. This degradation can be sped up or slowed down. Movement of an mRNA transcript out of the nucleus into the cytoplasm can be blocked or slowed down.

**RNA interference** can stop translation. Small strands of RNA (microRNA, miRNA) are formed when an mRNA transcript folds forming a double stranded structure. The loop is enzymatically removed; one strand of the resulting RNA is degraded; and the remaining single strand of miRNA combines with a protein. This protein/miRNA complex can complementarily bind to any mRNA and either block translation or degrade the mRNA. Small RNAs have been implicated in other regulatory activities as well.

When viruses attack cells, if their viral genome is a double stranded RNA, a process similar to that which produces miRNA can occur and the viral RNA/protein complex can breakdown production of proteins by the host cell.

Gene control is also responsible for development of an overall body plan known as **morphogenesis**. How an organism develops a front/back and anterior/posterior is the result of unequal distribution of mRNA, proteins, and organelles produced by the mother and deposited into an unfertilized egg. For example, in fruit flies, where mRNA transcribed from the maternal bicoid gene is concentrated in the egg, the anterior end of the animal will develop. Conversely, where mRNA from the maternal nanos gene is concentrated will become the posterior end of the organism once fertilization occurs. These maternal contributions that determine body plan formation are called cytoplasmic determinants.

After fertilization, the egg will undergo many cell divisions with each nucleus a replica of the initial zygote nucleus. However, as cytokinesis subdivides the initial egg cytoplasm with its unequal distribution of mRNA and proteins, each cell cytoplasm will have different mixes of the maternal cytoplasmic determinants. These will influence what genes within that particular cell's nucleus will be expressed, resulting in cell differentiation. Once differentiation occurs, the cell starts expressing particular genes and other genes are turned off, usually permanently. As the cell expresses its genes, it releases signaling molecules, and these influence the gene expression of neighboring cells in a process called **induction**. This lays the groundwork for body plan formation and for differentiation of cells, each functioning within its correct location.

The genes that specify body parts and location are **homeotic genes** and are highly conserved. This means they are critical to an organism's survival. Each homeotic gene has a similar stretch of DNA within it called a homeobox (a.k.a. Hox).

> **DID YOU KNOW?**
>
> Epigenetics is the study of heritable changes that occur when chemical reactions affect the expression of genes. For example, many genes are methylated which usually silences the gene—it is not expressed. However, the epigenome is flexible and can respond to environmental cues. This means that a particular gene may be present but never expressed, or expressed as a result of changes to these external, nongenetic control factors. Two genetically identical mice, for example, fed two different diets, one with normal amounts of the raw materials for methyl production and one very low in these raw materials show very different phenotypes. The mouse with the deficient diet is obese, yellow colored, and diabetic, but the mouse not deficient is sleek, brown, and healthy. If these adult rats are fed normal diets, the obese mouse will give birth to healthy pups! The genome was not changed, the expression of the genome changed. BPA (Bisphenol A) a component of plastics decreases methylation and healthy mice exposed to BPA show the same health issues as the obese, sickly, yellow mouse.

Cancer can result when gene control fails. The tumors that result from cancers contain cells that differ from their normal counterparts in many ways and these cells are not able to perform their normal functions. The particular types of genes that are susceptible to initiating cancer are **proto-oncogenes** and **tumor suppressor genes**.

Proto-oncogenes are normal genes that code for proteins that function in cell growth and division. Proto-oncogenes become oncogenes when a change causes over-production of its protein product or the protein that is produced is impaired. An example of an oncogene is the *Ras* protein, which normally initiates a phosphorylation cascade that ends with production of a protein that increases cell division. If the *Ras* gene is mutated, the resultant protein continually triggers the pathway in the absence of any signal. The result is increased cell division. Most oncogenes are dominant.

Tumor suppressor genes usually function in control of the cell cycle. They encode proteins that inhibit cell division, enzymes that function in DNA repair, or proteins that promote apoptosis. An example is the *p53* gene. This gene encodes proteins that inhibit the cell cycle, activates DNA to produce DNA repair enzymes, and can initiate apoptosis. Tumor suppressor genes generally act as recessive genes.

# Applications of Genetic Engineering

**Genetic engineering** is the manipulation of genetic material to ultimately make useful products through **biotechnology**. It took an understanding of DNA, the molecular concept of a gene, and gene regulation, as well as the discovery of molecular "tools," before scientists could effectively manipulate genetic material. Some of the molecular "tools" were normal molecules found in microorganisms. The most important molecular tools are:

1. **Plasmids** are small extranuclear circles of DNA found in bacteria and yeast. These carry nonvital but potentially useful genetic material, and they are replicated along with the cell's main chromosome. Plasmids can be used as **vectors**, carriers of genetic information.

2. **Restriction endonucleases** are enzymes that cut up nucleic acids. Bacteria protect themselves from viral infection by using restriction enzymes to cut up the viral genome that enters the cell. Restriction enzymes cut DNA at very specific sequences and the cut DNA may be blunt or staggered. The useful cut is staggered because it yields sticky ends—"tails" of single stranded DNA to which a complementary strand will readily attach. For example, EcoRI is a restriction enzyme that makes a staggered cut between G and A bases in a palindromic sequence:

```
5'  G | A   A   T   T   C  3'
3'  C   T   T   A   A | G  5'
```

3. **Ligase**, DNA polymerases, and RNA polymerases are all naturally occurring enzymes used in making copies of genetic material. Specifically, ligase is an enzyme which covalently bonds DNA pieces together.

4. **Reverse transcriptase**, an enzyme produced by retroviruses, allows scientists to produce DNA from RNA. This has proven useful in eukaryotes. mRNA from an active cell can be transcribed into **cDNA,** eukaryotic DNA without introns. This cDNA can then be used as source DNA for recombination.

5. The discovery of **thermophilic** bacteria with heat tolerant DNA polymerase enabled scientists to automate amplification (cloning) of DNA.

## Genetic Engineering Techniques

**Cloning** DNA is a first step in biotechnology because it is the process of making many genetically identical copies of a particular gene, segment of DNA, or individual organism of interest. Cloning a gene or DNA segment can be done several ways. The most efficient way is through **recombinant DNA**, the combining of DNA segments from two different, usually unrelated, organisms. The steps are:

1. Identify and isolate the desired DNA.

2. Select a plasmid to use as a vector.

3. Cut the source DNA and the plasmid vector with the same restriction endonuclease to yield complementary sticky ends.

4. Mix the cut plasmid and source DNA along with ligase. Some of the cut plasmids will combine with the source DNA and be recombinant plasmids.

5. Introduce the vector into bacterial cells.

6. Allow the bacterial cells to multiply on a medium that will select for **transformants**.

Recombinant DNA technology can be used to establish a genomic library from which any particular piece of DNA can be isolated using hybridization. A stretch of the desired DNA must be known in order to make a complementary single-strand DNA probe. The probe is systematically exposed to the library of DNA where it will base-pair with the DNA clone carrying the gene of interest.

**Bacterial transformation with cDNA** from eukaryotes can be used to produce large quantities of pharmaceutical products. Insulin and human growth hormone are two important products of this biotechnology.

The **polymerase chain reaction, PCR**, is another way of making many copies of a section of DNA. A piece of DNA is mixed with nucleotides and enzymes necessary for replication. Key to successful PCR is selecting primers that will delineate the target sequence so that it is disproportionately replicated in subsequent rounds of replication. The mixture is heated to denature the DNA sample, primers are added, and the mixture is cooled so that the primers will bond to the target sites. Polymerases will add nucleotides to the 3´ end of each primer. This cycle of heating, cooling, and replication is repeated many times in order to make sufficient copies of the target DNA. PCR can be used to make clones of even very small samples of DNA. This technique is often used in forensics or diagnosis as a precursor to **gel electrophoresis**.

**Gel electrophoresis** is a method of separating pieces of DNA according to size. DNA samples are cut with specific restriction enzymes. The resulting pieces, restriction fragments, are placed in wells formed in a gel that is usually agarose, a polysaccharide extract of agar. Agarose forms a matrix and, when a charge is placed across it, the DNA pieces will move through the gel matrix away from the negative pole towards the positive pole. The agarose impedes the progress of the DNA pieces so the smaller pieces move farther through the gel than the larger pieces given the same amount of time and charge. The bands can be visualized and measured. Gel electrophoresis is useful for comparing pieces of DNA. For example, certain DNA samples from two different humans cut with the same restriction endonucleases will not be the same size so they will show different banding patterns. For identifying criminals, areas in the DNA where there is variation in **short tandem repeats**, segments of DNA that repeat a variable number of times, are used for gel comparisons.

The DNA of many organisms, including humans, has been sequenced. This information can be used to compare genetic information from different organisms. Sequencing requires source DNA, a primer that is complementary to the 3´ end of the source DNA, replication enzymes, regular nucleotides, and specialized fluorescent nucleotides. The specialized nucleotides fluoresce a different color for each nucleotide base.

The sample DNA is denatured, the primer added, and synthesis of the DNA proceeds. Each time a specialized nucleotide is placed complementarily, replication ends on that strand. This will result in a series of strands of DNA of different lengths that fluoresce their particular nucleotide's color at the end. These strands are then read from shortest to longest.

## Applications of Genetic Engineering

Applications of DNA technology include forensics, diagnosis of disease or identification of carriers, production of useful pharmaceuticals, and studies of gene expression and evolutionary patterns. Some areas of biotechnology pose questions about appropriate and ethical use. Genetically modified crops may make a plant healthier and more resistant to pests; however, a crop is not isolated from its surroundings and the genetic modification may find its way into wild plants creating super weeds. On the other hand, golden rice is genetically modified to synthesize beta-carotene, a precursor to vitamin A. Vitamin A deficiency is the leading cause of preventable blindness, so such a product is a dietary boon to areas of the world with rice-dependent diets. Animals can be cloned, but because of epigenetics, it is not possible to completely control the results of the cloning.

> **DID YOU KNOW?**
>
> Restriction endonucleases are named after the organisms from which they were taken and in the order of their discovery within that organism. For example, *EcoRI* is a restriction endonuclease from *Escherichia coli* bacteria and was the first one identified. *SmaI* comes from *Serratia marcescens* and makes a blunt cut between the C and G of the palindromic sequence 5'CCCGGG3'. Most bacteria have many different restriction endonucleases to use in defense against bacteriophages.

## Genotype Affects Phenotype

Unfortunately, the process of DNA replication doesn't always work perfectly and mistakes can occur in the DNA sequence. A change in the DNA nucleotide sequence of an organism is a **mutation**.

Both internal and external factors can cause a mutation. **Internal factors** include a mistake in DNA replication or in repairing DNA. **External factors** include exposure to mutagenic chemicals, X-rays, and UV sunlight.

Mutations can have a positive, negative, or no effect on an organism's phenotype. The effect depends on whether or not the mutation affected the protein product made in translation. For example, sickle-cell anemia is caused by a change in a single nucleotide, causing hemoglobin to change shape. Other mutations will be silent because they occur in noncoding DNA or in the third nucleotide in a codon and produce the same protein product.

A mutation that causes an alternation of the transcribed protein is deemed positive or negative based upon the environment of the organism. For example, the altered hemoglobin in sickle-cell anemia confers the positive benefit of increased resistance to malaria. Thus, an individual having one gene for sickled hemoglobin (heterozygous) and living in a tropical area has a selective advantage over a person without sickled hemoglobin (homozygous); this phenomenon is called heterozygote advantage.

## TEST TIP

Although memorizing specific genetic diseases isn't required for the AP Biology exam, understanding heterozygote advantage is a good idea because several human genetic disorders are due to this. Thalassemias, sickle-cell anemia, and cystic fibrosis are all disorders in which having one "bad" allele conferred an evolutionary advantage. Thus, the alleles remained in the gene pool.

Changes in phenotype also can occur due to changes at the chromosomal level. The chromosomal number can be altered due to nondisjunction events in mitosis or meiosis if sister chromatids or homologous chromosomes do not segregate properly.

**Aneuploidy** occurs when an individual has an abnormal number of chromosomes. A cell having one extra chromosome is called trisomic, while missing one chromosome is called monosomic.

**Polyploidy** involves having more than two sets of chromosomes; for example, triploid (3n) instead of diploid (2n). Polyploidy has less of a phenotypic effect on an organism than aneuploidy; one extra chromosome upsets the genetic balance more than an extra set of chromosomes.

In humans, aneuploidy is usually lethal, with a few exceptions such as trisomy 21 (Down syndrome) in which the individual has developmental problems. Polyploidy has not been observed in higher mammals like primates.

In plants, having extra sets of chromosomes is not problematic and is even common. For example, wheat is hexaploid (6n) and strawberries are octaploid (8n). Polyploidy gave rise to the living organism with the most chromosomes: an adder's tongue fern has over 1,200 chromosomes.

Essentially, changes in genotype that affect phenotype can affect a population's gene pool and lead to evolution of a population. Thus, through natural selection and environmental pressures, genotypic alternations that are beneficial in an organism's ecosystem will be preserved in the gene pool.

Mistakes occurring in the process of DNA replication or DNA repair lead to variation in the genotypes of individuals. Additional genetic variation is acquired through sexual reproduction in eukaryotes. Even though they reproduce asexually, prokaryotes have methods of exchanging DNA to increase variety.

Prokaryotes have two primary methods of obtaining new genetic material. This is called horizontal gene transfer. **Transformation** involves the uptake of DNA that is found freely in the environment. **Transduction** occurs when a virus transmits new genetic material to a bacterial cell.

Eukaryotic sexual reproduction has several methods that increase variation. First is the formation of haploid gametes that will recombine with another gamete—essentially, offspring are a combination of two different sets of DNA. The processes of crossing-over and independent assortment during meiosis also increase variety.

## Viral Replication

Why are viruses included in a chapter about genetic variation? Essentially, viruses are infectious packages of DNA or RNA in a protein shell called a capsid.

Viruses infect host cells and use the host cell machinery to create more viruses. The method of viral reproduction can introduce foreign genetic material into the host cell, leading to a change in the genome of the host. For example, many cancerous cells have been found to have viral DNA in them, such as the immortal cell line from the patient Henrietta Lacks (HeLa cells), which have DNA from the human papilloma virus.

Viral genetic material can come in double- or single-stranded DNA or double- or single-stranded RNA. There are three basic mechanisms used by viruses to reproduce. They are the **lytic cycle**, the **lysogenic cycle**, and **reverse transcription**.

1. Viruses that use the **lytic cycle** have a five-step process of infection to release new virus particles. First, the virus attaches to specific receptors on the host cell surface. Second, the viral genetic material is injected into the cell and typically the host cell's genetic material is destroyed. The viral DNA takes over the cell's enzymes and ribosomes to synthesize viral DNA and capsid proteins. The viruses assemble and finally are released from the cell, killing the host cell in the process.

2. The **lysogenic cycle** does not kill the host cell. Instead, the viral genome integrates into the host genome. The first steps of this cycle involve attachment of the virus to the host cell and insertion of the viral genome into the host cell. Unlike the lytic cycle, the viral DNA will be incorporated into the host cell's DNA, forming a **prophage**. The prophage DNA remains silent and is copied along with the host cell's DNA every time the host cell divides. Thus, every daughter cell will also contain a copy of the prophage DNA. Expression of prophage genes can change the host cell's phenotype; for example, the botulinum toxin produced by some strains of the bacteria *Clostridium botulinum* is created from viral prophage DNA.

   Essentially, many of these viruses, such as the lambda bacteriophage, can alternate between the active lytic cycle and the dormant lysogenic phase. Typically, an environmental trigger causes the prophage DNA to leave the host genome and take over the cell's replication machinery.

3. Retroviruses are RNA viruses that use **reverse transcription** to replicate the viral genome and proteins. The most commonly used example of a retrovirus is HIV (human immunodeficiency virus), an enveloped single-stranded RNA virus. The first step of infection by a retrovirus binds the viral envelope proteins to receptors on a host cell. The viral proteins and RNA are then released into the cell. One of these proteins is a special enzyme called reverse transcriptase (RT) which transcribes the viral RNA into DNA. RT will transcribe the RNA into two strands of DNA, which incorporates into the cell's DNA, forming a **provirus**. The proviral DNA will be transcribed and translated, creating the RNA and proteins needed to make new virus particles, which will ultimately bud off the host cell. They do not kill the host cell, which allows more viruses to be continuously made.

Because viruses are using the cellular machinery from their hosts, **mutations** arise in viral genomes just as they do during prokaryotic and eukaryotic cell replication. RNA viruses have an even higher rate of mutation than DNA viruses because they do not use the error-checking mechanisms that are used in DNA replication machinery. Due to the speed and volume of viral replication, mutations can accumulate quickly in viruses leading to the rapid evolution of the virus. This is one of the reasons why treating viral infections is so difficult from a medical standpoint. HIV is an excellent example of this; it is an RNA virus that mutates rapidly, making it complicated to treat or prevent.

## Questions to Consider

1. What might be the consequence to a cell if a virus introduced an enhancer into a region of DNA?

2. If you took fruit fly homeobox sequences, denatured the DNA, mixed it with denatured DNA from other organisms, then allowed cooling, what might you find?

3. Trace a mutation in DNA to its phenotypic outcome.

4. Thinking in terms of evolution and changes to genetic material, explain why there is a new flu vaccine every year.

## Answers:

1. That region of DNA would be subject to being transcribed more frequently or transcribed under incorrect action signals. This would have the potential of starting cancer if this was in an area of DNA that coded for cell division proteins.

2. Every animal tested would have complementary homeobox sequences, indicating that virtually all animals share these ancient, conserved genes.

3. Let's say the enzyme to digest lactose contains a TTC (mRNA – AAG) codon that is mutated to a CTC (mRNA – GAG). This would change the amino acid that goes into that protein from lysine to glutamic acid. Lysine is a positively charged, basic amino acid while glutamic acid is acidic and negatively charged. This is a significant change in amino acid type and would cause the three-dimensional shape of the protein product to change. This change in shape is likely to mean the enzyme will no longer function and that person will no longer be able to digest lactose. This would affect the person's phenotype and make him lactose intolerant. (Please note: these sequences were made up for this problem.)

4. There are many strains of flu viruses, and each is constantly changing and accumulating mutations. Thus, the strain of flu that may be most prevalent one winter can mutate so much that its DNA is very different by the next winter, thus requiring a new vaccine to target the newly evolved strain. Additionally, the strain of flu that is prevalent one year may not be the next year.

# Chapter 17

# Cell Communication

## Cell Communication

Cells do not operate in a vacuum. Instead, they must maintain homeostasis with the environment around them. This involves creating, transmitting, and receiving signals from other cells and from the surrounding environment. These signals can either stimulate a response, such as increasing gene expression of a needed protein, or inhibit a response, such as factors that prevent cell growth, resulting in density dependent inhibition. Despite the vast differences in unicellular and multicellular organisms, both use similar signal transduction pathways of receiving a signal, transduction of the signal down a chemical pathway, and response to the signal. The similarity of the pathways indicates a shared evolutionary past.

Unicellular organisms use chemicals and signal transduction pathways to appropriately respond to the environment around them. Environmental cues can include chemical messengers from nearby cells that trigger movement or growth. Quorum sensing in which bacteria regulate the density of their population is one example. Bacteria send out small chemical signals that are picked up by cells in the vicinity, ultimately allowing the group of cells to coordinate their growth and cellular functions. In another example, pheromones released from other cells can trigger reproduction. Yeast of mating type *a* secrete an *a* chemical signal that binds to receptors on type α cells, while type α cells secrete an α signal that actives *a* cells. As a result of the signaling pathway initiated by binding of the chemical messenger, the two different types of cells will grow towards one another and mate, resulting in genetically new offspring.

Multicellular organisms use similar signaling pathways to regulate their more complex systems. These organisms have many differentiated cells with differing functions; cell-to-cell communication allows the coordination of many individual cells into a working tissue, organ, or organ system. Ultimately, their signaling pathways direct physiological responses, complex behaviors, and response to the environment. An important signaling molecule in animals is epinephrine, also known as adrenaline, which has a multitude of effects on tissues as well as behavior. It will be used to outline a signal transduction pathway involving reception, transduction, and response.

> **DID YOU KNOW?**
>
> The bacteria *Vibrio fischeri* can be found living inside squid and are capable of giving off a greenish light, making the squid appear to glow. *V. fischeri* will not glow when they are scattered in ocean water because their population density is too low. However, when concentrated inside the squid, they will glow because their density is high. This system works through signal transduction. The bacteria give off a signaling molecule into their surroundings. If there are many cells nearby to receive the signal, as in the squid, then each cell will glow.

1. Reception of the hormone epinephrine occurs when it binds to adrenergic receptors on a target cell in the body. Adrenergic receptors are a type of G-protein-coupled receptors.

2. When these receptors are activated, they activate a signal transduction pathway which begins by increasing the concentration of cAMP (a second messenger). cAMP triggers the activation of protein kinase A, which sets off a phosphorylation cascade.

3. The cell type determines the exact response to the phosphorylation cascade. In a liver cell, the final protein activated causes the breakdown of glycogen and releases glucose.

Essentially, through a complex series of reactions, epinephrine stimulates the release of glucose from liver cells. If any one part of this complex pathway is disrupted, glucose will not be released.

# Chemical Signaling

Chemical signals are one of the ways that cells communicate with each other. This can occur through direct contact or over short and long distances.

**Direct contact** occurs when one cell is in physical contact with another cell. Typically in cell-cell recognition, proteins in the cell membranes come into contact with one another and trigger or inhibit a signal. An example occurs in the immune system when an infected cell displays a portion of the foreign antigen on an extracellular protein. A helper T cell will recognize and bind to this display, signaling the infected cell to release cytokines, which in turn signal a larger immune response. Another example is the gap junctions in animal cells or the plasmodesmata in plant cells that act as tunnels between cells, allowing materials to flow directly between cells that are in physical contact.

**Local signaling** occurs over short distances and involves the release of a messenger molecule being secreted by one cell. These messenger molecules are called local regulators and affect cells that are in close proximity to the secreting cell. There are two types of local signaling: **paracrine** and **synaptic signaling**.

- An example of paracrine signaling is growth factors whose release may trigger the growth of cells in the vicinity of the cell secreting the factor.

- Neurotransmitters in animal cells are an example of synaptic signaling. When an electrical signal reaches the end of a neuron, it triggers the release of a chemical neurotransmitter into the synaptic space between two neurons. When the target cell receives the chemical message, it will respond appropriately.

**Long-distance signaling** occurs through **endocrine signaling**, which is the release of a hormone by a specialized endocrine system into the blood stream. The hormone will act on a target cell having the appropriate receptor. Plants also use hormones, although they are often gases (such as ethylene which triggers fruit ripening) that can diffuse from cell to cell or through the air. One animal example is the use of steroid hormones (such as testosterone or estrogen), which bind to intracellular receptors. Many of these activated hormone-receptor complexes act as transcription factors that regulate gene expression. A second example of endocrine signaling is the use of insulin to control glucose levels in animals. The release of insulin from the pancreas into the bloodstream triggers body cells to remove glucose from the blood.

# Signal Transduction

**Signal transduction** occurs when a signal is initiated by a chemical messenger, called a **ligand**, and then is received by a receptor. The ligand and receptor are specific for each other; a receptor can only be bound by one ligand. When a ligand binds, it induces a shape change in its receptor, which will initiate signal transduction. The receptor is a protein either embedded in the membrane or found within the cell.

**Membrane-bound receptors** usually bind to large, polar ligands, such as epinephrine. Common examples include G-protein-coupled receptors, receptor tyrosine kinases, and ligand-gated ion channels. More specifically, when a ligand binds to a G-protein-couple receptor, the receptor changes shape. This shape change allows the receptor to bind to a G-protein, which is found inside the cell. The G-protein becomes activated, dissociates from the receptor, and binds to an enzyme, ultimately activating that enzyme and setting off a cellular signal.

**Intracellular receptors** usually bind small, non-polar molecules, such as testosterone, or gases, such as nitric oxide. For testosterone, the steroid hormone diffuses across the cell membrane and into the cytosol where it binds to a receptor protein. Binding activates the receptor protein, causing it to move into the nucleus and bind to specific genes. Thus, this hormone-receptor complex acts as a transcription factor to regulate gene expression.

## TEST TIP

Memorizing specific pathways, such as the G-protein-coupled receptors and receptor tyrosine kinases, is not required for the exam. However, it would be wise to choose one of the systems and learn it as an illustrative example you could discuss if it appeared in a free-response question.

Once the signal is received, it must lead to a cellular response. Transduction typically involves a cascade of chemical changes, often to proteins, that eventually lead to the response. A phosphorylation cascade is a common example. Phosphorylation occurs when a phosphate group is added to a molecule and is accomplished by enzymes called protein kinases. In a phosphorylation cascade, once the receptor is activated by its ligand, it will trigger a molecule that activates a protein kinase; this protein kinase will phosphorylate a second protein kinase, and so on, until the final molecule is activated and causes the cell to respond.

Another way to initiate a cellular response is to continue the signal using a **second messenger**, small molecules that easily spread through a cell. The most common examples are cyclic AMP (cAMP) and calcium ions ($Ca^{2+}$). An activated G protein causes an enzyme called adenylyl cyclase to convert ATP to cAMP. Usually cAMP activates protein kinase A, which will set off a signaling cascade and cause a cellular response. $Ca^{2+}$ is used more often than cAMP and is common in muscle cells and neurons. It is released from the endoplasmic reticulum, rapidly diffuses through the cell activating proteins, causing a cellular response.

So why does a cell bother with such complex pathways instead of a direct signal? One of the biggest advantages to signaling cascades is the ability to significantly amplify the response. For example, when one epinephrine molecule binds to a G-protein-coupled receptor on a liver cell, it can cause the activation of approximately 100 G-proteins, leading to the release of almost 100,000,000 glucose molecules.

One of the emerging fields of medical research is due to the recognition that disruptions to these signaling pathways can either prevent or over-activate cellular response, leading to disease or even death. The disease cholera is caused by a bacterium, *Vibrio cholerae*. The bacteria secrets a toxin in the small intestines; the toxin modifies a G-protein that regulates salt and water concentrations, leaving it continuously active. This produces a high concentration of cAMP, which causes the intestinal cells to secrete salt. To maintain osmotic balance, water will diffuse from the cells to follow the salts, leading to profuse diarrhea and dehydration in a person infected with the bacteria. Cholera regularly affects and often causes the deaths of people in developing nations who do not have access to clean drinking water.

Additionally, we have learned to block these pathways with the use of prophylactic drug treatment. One familiar example is the antihistamines found in cold and allergy medicines. In an allergic reaction, mast cells of the immune system, which are found in mucosal membranes, bind to foreign antigens such as pollen. This binding sets off a signal transduction pathway, ultimately releasing histamines, which cause watering of the eyes and a runny nose. Antihistamines in medication function by blocking the histamine receptors, ultimately stopping the signal cascade and preventing the symptoms of the allergic reaction.

## Questions to Consider

1. Explain what is meant by the terms *reception, transduction*, and *response* and why all cells must do these steps in order to survive.

2. Read the following description of the hormones involved in regulation of the menstrual cycle.

   In females beginning a normal menstrual cycle, estrogen levels increase until they peak at about day 14. The release of estrogen causes an increase in luteinizing hormone (LH) and follicle stimulating hormone (FSH), causing the release of an egg from a mature follicle in one ovary. Ovulation triggers the rise of progesterone levels. If the egg is not fertilized and embryo implantation in the uterine wall does not occur, estrogen and progesterone levels drop significantly, causing the shedding of the uterine wall. Birth control pills work by using synthetic hormones to maintain constant levels of hormones.

   Explain how this is an example of a signal transduction pathway and suggest how birth control pills work to prevent ovulation.

## Answers:

1. Reception occurs when a ligand binds to a protein receptor. Transduction occurs when the activated receptor sets off a pathway that continues to send the signal, often amplifying the signal in the process. Response occurs when the transduction pathway ends and causes the cell to respond in some way, such as by making a protein product or breaking down glycogen into glucose. These steps are vital for all cells because cells must be able to communicate with one another and receive signals from their environment.

2. The menstrual cycle is regulated by a specific set of hormones. The timing of their release and the amount of hormone present is critical. The cycle begins with estrogen levels increasing, which triggers a surge of LH and FSH. This surge causes the release of the egg and increases in progesterone. By maintaining constant levels of hormones, the signaling pathway is disrupted, thus preventing egg release.

# Chapter 18

# Information Transmission

## Information Exchange

Organisms receive and process information from their environment and from other living organisms using their sense organs and nervous systems. Being able to communicate and process this information is vital for the survival of both individuals and populations of organisms because animals that respond appropriately to external signals are more likely to survive. For example, vervet monkeys have three specific predators and use alarm calls specific to each predator. Thus, if a monkey hears the alarm call for a leopard, it will climb a tree and perch on a branch far from the trunk in order to avoid the leopard.

Many organisms use behaviors, visual signals, or chemical cues that cause other organisms to modify their actions in a specific way, ultimately leading to the reproductive success of the first organism. For example, fruits turn bright colors to indicate ripeness. This color cue indicates to animals the fruit can be eaten, in the process distributing the seeds of the plant away from the parent. This distribution of seeds decreases competition between the parent and offspring and makes survival of those genes more likely. Another example is an animal scent marking its territory, often with urine, in order to let other animals know that a particular territory is already occupied. As a result, scent marking can decrease competition for resources and mates as long as each animal's territory continues to provide ample food and mating choices.

Essentially, animals use five means to communicate information to one another: visual, tactile (touch), auditory, chemical (pheromones), and electrical. The table below gives an example of each type of communication.

Table 18-1. Methods of Animal Communication

| Method of Communication | Example |
| --- | --- |
| Auditory | Many male birds sing to attract mates or signify territory. |
| Chemical | A female moth secretes pheromones into the air to indicate she is ready to reproduce. |
| Electrical | Some fish have organs that generate a weak electric field. This is used to attract mates or to signal territory. |
| Tactile | Before mating, male fruit flies tap female fruit flies with one of their legs to signal their presence. |
| Visual | Bees, yellow jackets, and wasps have yellow and black coloration to warn potential predators that they are harmful. |

Each of the behaviors leads to increased fitness in the organisms communicating. Using the female moth secreting pheromones from Table 18-1 as an example, a male moth who can receive this signal is more likely to mate and pass its genes on to a future generation.

**Altruistic behavior** is a type of social communication that also leads to the evolutionary success of a population. Altruistic behavior is when an animal behaves in a way that decreases its own fitness in order to increase the fitness of the overall population. For example, bees communicate food locations by "dancing," a form of tactile communication. A bee colony that is better able to find and communicate food sources is more likely to survive than a colony that doesn't communicate food location. Another example is found in the herding behavior of prey animals or the schooling of fish. In both cases, animals group together in large numbers to make it more difficult for a predator to catch any single animal. Thus, populations of animals that herd or school are more likely to have higher survival rates. If an animal is missing the gene that causes the innate behavior of massing into the herd, it is more likely to die and its genes will be removed from the population's gene pool.

# Cognition

**Cognition** is the ability of the nervous system to perceive, store, process, and respond to information obtained by the senses. A cognitive map is an internal representation or code of the spatial relationships among objects in an organism's environment. Animals use cognitive maps to help them migrate, for example. Bees use cognitive mapping to remember and communicate the locations of food sources. Essentially, an animal's ability to perceive, store, process, and respond to its environment is crucial to its survival and to the overall well-being of the population. In fact, consciousness also plays a factor; consciousness is the awareness of one's self and the environment and requires cognition.

The nervous system works to transmit electrical and chemical signals to make cognition possible. The functional unit of the nervous system is a cell called a **neuron**. The structure of a typical neuron is shown in Figure 18-1.

Figure 18-1.  A Typical Neuron

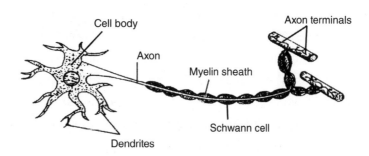

The neuron is an excellent example of how form fits function in biology because it is shaped to maximize detection and transmission of signals.

- The nucleus is found in the cell body (sometimes called the soma).

- Dendrites are projections off the cell body that act like antenna and pick up signals from neighboring neurons.

- The axon is the long, thin portion of the neuron down which nerve signals are sent.

- The axon is insulated by glia cells, which also protect the nerve cell. In the central nervous system, this proactive insulator is called the myelin sheath while in the peripheral nervous system, these cells are referred to as Schwann cells. There are gaps in the insulation, called the nodes of Ranvier, which allow for faster action potential—that is, faster communication among the nerve cells.

- The end of the axon branches into terminals, which release neurotransmitters in the **synaptic cleft**, the space between two neurons, as well as between the neuron and the target cells, such as muscle cells; this area is called the **synapse**.

Nerve signals are transmitted through both electrical and chemical methods. Electrically, neurons use differences between intracellular and extracellular concentrations of sodium ($Na^+$) and ($K^+$) ions to establish gradients in charge, which leads to a voltage difference on either side of the cell membrane. This voltage difference is a source of potential energy and is called **membrane potential**. At rest, a neuron has a negative membrane potential, which is established by sodium potassium pumps in the cell membrane. Using energy from ATP, these membrane protein pumps actively transport two $K^+$ ions inside the cell and three $Na^+$ ions outside the cell.

If a neuron receives a stimulus, gated $K^+$ and $N^+$ channels will open, allowing ions to flow down their chemical gradient and causing a change in the membrane potential. Due to the movement of these ions, the membrane potential will become more positive, which is called depolarization. If the depolarization is positive enough, called the threshold, then the neuron will propagate an **action potential**. This is a quick change in membrane voltage that travels down the axon of a neuron. Once an action potential is started, it cannot be stopped. Immediately after the depolarization, the ion channel gates close, to causing repolarization; there is a slight delay after each ion channel gate closes to encourage the action potential to move only in one direction and not to reverse itself. The one-directional movement of the action potential away from the cell body and towards the synapse is crucial to helping neuron cell signaling and communication with other neurons.

Once the action potential reaches the synapse of the neuron, chemical transmission commences. **Chemical transmission** involves sending impulses between neurons across the space between two neurons, which is called the synaptic cleft. When an action potential reaches the end of an axon, the terminal, it depolarizes voltage-gated channels that allow calcium to flow into the axon. The increase in calcium causes synaptic vesicles, which contain chemical neurotransmitters, to fuse with the cell membrane and release the neurotransmitter into the synapse. The neurotransmitter diffuses across the small synaptic space to the next neuron (called postsynaptic), which has receptors in its membrane. The response of the postsynaptic neuron depends on the neuron type, the neurotransmitter, and the receptors in the postsynaptic membrane. If it causes depolarization, then it generates a response and is called excitatory postsynaptic potential. If it causes hyperpolarization, it will prevent a response and is called inhibitory postsynaptic potential. Thus, a neurotransmitter can either stimulate or inhibit a response.

In summary, the path of transmission of information between neurons is:

Dendrite → cell body → axon → axon terminal → synaptic cleft → dendrite of next neuron

There are many important neurotransmitters used in animals, and scientists have learned that many nervous system disorders (from depression to Parkinson's disease) are due to imbalances of these transmitters. An example of a neurotransmitter is epinephrine, sometimes called adrenaline, which acts on almost all body tissues. For example, epinephrine is released into the bloodstream when a person becomes frightened. It is often called the "fight or flight" response because it increases the heart and breathing rates, increases blood glucose concentrations, and stimulates muscle contractions, preparing that person for immediate action.

The vertebrate nervous system is composed of peripheral nerves, sensory organs, the spinal cord, and the brain. The brain is made of billions of neurons massed together sending and receiving signals through many neurotransmitters. The brain is divided into sections, each of which has a specialized function. For example, the frontal lobe functions in decision making, controlling skeletal muscle, and speech, while the temporal lobe functions in hearing and understanding language. The brain and spinal cord consist of gray matter and white matter. White matter is found in the brain's interior and the spinal cord. It consists of myelinated axons, which is consistent with its function in learning, rapidly processing information from the peripheral nervous system (PNS), creating emotions, making decisions, and producing commands for the PNS. Additionally, the white matter transmits impulses between the PNS, the spinal cord, and the brain.

# TEST TIP

Memorizing the parts of the brain is not necessary for the AP Biology exam.

# Regulation of Behavior

Organisms must time and regulate their behaviors with each other in order to survive environmental conditions, reproduce, and pass their genes on to their offspring. The first step is for organisms to communicate information with one another; then they must act appropriately on the information received from others. In animals, behaviors are the responses to the external signals they receive, and there are two types: innate and learned.

Both innate behaviors (ones that are instinctual) and learned behaviors help an organism survive to reproductive age.

- **Innate behaviors** are mostly determined by genetics and are developmentally fixed and unchanging; innate behavior is sometimes referred to as "instinct." The simplest type of innate behavior is a **fixed action pattern**, a sequence of actions that occur in the same pattern anytime an animal experiences the stimulus. An example occurs in a male stickleback fish, which has a red abdomen, and will lunge and attack any other object having a red underside that invades its territory, including plastic objects.

- **Learned behaviors** are variable and occur after an organism interacts with its environment or another organism; thus, they are a result of that organism's experiences. One common example is spatial learning. Every organism's habitat has differences in the layout of dominant plant species, sources of water and food, possible shelter, and the location of potential mates and predators. An animal that can form a spatial memory of the environment learns to navigate the environment and is better able to conserve energy to find food and water and shelter, locate mates, and avoid predation.

> **DIDYOUKNOW?**
>
> It is not only mammals that learn. In an experiment from the 1950s, Stuart Sutherland taught octopuses how to distinguish between a square, a horizontal rectangle, and a vertical rectangle.

Essentially, an animal's ability to respond to environmental signals and communicate with other organisms determines its evolutionary fitness. Those organisms that are incapable of appropriate responses will be selected against. One example of an innate behavior influenced by natural selection is **migration**, the movement of animals from one location to another, typically over a long distance, in response to an environmental cue. Birds that "fly south" for the winter are one type of migrating animal. A bird whose genes do not cause it to recognize environmental cues to the changing seasons will not migrate and is likely to die of starvation or cold. This removes its genes from the gene pool; thus, the birds that are genetically programmed to seek warmer climates for the winter will be fitter and survive.

Additionally, organisms within a population that exhibit **cooperative behaviors** (either within the same species or with another species) increase the chances of survival of both the individual as well as the entire population. The existence of the large number of mutualistic symbiotic relationships on Earth attests to the importance of cooperation. The relationships vary amongst the kingdoms and include:

- Protists living in the digestive system of termites—Termites cannot digest cellulose in wood, but the protists can. Thus, both organisms receive energy.

- The relationship between fungi and algae in forming lichens—The fungus provides rooted stability and the alga can photosynthesize, producing energy.

- Bacteria living in the digestive tracts of ruminant animals—Cows cannot digest cellulose; however, the bacteria can. Both organisms receive energy.

- Sea anemones and clownfish—The anemone protects the clownfish from larger predators. The clownfish protects the anemone from its predator, and clownfish waste is food for the anemone.

- Oxpecker birds and many types of cattle—The birds eat parasites, such as lice or ticks, off the cattle. This provides the birds with food and keeps the cattle parasite-free.

- Flowering plants and pollinators—The pollinator gets food (either nectar or the pollen itself) from the flower while the plant gets its eggs fertilized. The pollinator will also spread the pollen (which contains the sperm) to other plants, increasing genetic variability in that plant species. This is called cross-pollination.

## Questions to Consider

1. A vervet monkey sends an alarm call for a snake, and the monkeys nearby promptly stand up straight and look at the ground around them.

    (a) Explain how information is transferred within the population of monkeys.
    (b) Explain how information is transferred within a monkey who hears the alarm call.
    (c) Relate the behavior of the vervet monkeys to evolution.

2. How does the nervous system work with other systems in the body to allow an animal to walk?

3. Throughout North America and Europe, honey bees are dying on a mass scale in a phenomenon referred to as "colony collapse disorder." The exact cause of the bees' death is unknown. Several large-scale agricultural operations (for example, California's almond industry) are concerned about this disorder. Why might this be?

## Answers:

1. (a) One monkey spots the snake and makes the auditory call that signifies a snake has been seen. All the monkeys that are close to the calling monkey hear the auditory signal and recognize it as the signal for snake. Each monkey will look at the calling monkey first, then respond with the appropriate behavior. Only the monkeys that are within hearing distance will respond.

    (b) One monkey will hear the alarm call through the sense organ of the ear. This will activate a signal in its nervous system. The signal will travel from one neuron to the next, until it reaches the spinal cord, and finally the brain. The brain will register this as an alarm call and process that an action needs to be taken. The brain will then send a signal down another pathway of neurons that will eventually reach skeletal muscles and cause the monkey to move appropriately.

    (c) All monkeys have slightly different genetic makeups. Due to their genes, some have better hearing, some have better vision, and some have faster response times. A monkey with poor hearing may not hear an alarm call and get eaten, thus removing its genes from the gene pool. Similarly, a monkey with better vision is more likely to see the snake and avoid being eaten, thus leaving its genes for good vision in the population's gene pool.

2. The brain initiates a signal to neurons that the animal should walk. This signal starts in a neuron and travels from neuron to neuron until the signal reaches all the necessary muscles in the animal's legs. Both electrical and chemical messengers are used to deliver this signal to the muscles to begin contracting and relaxing, hence walking. It's important to remember that the nervous system, like many other systems in the body, works in concert with other systems and organs to allow movement, reproduction, health, and survival.

3. Honeybees and almond flowers are a prime example of cooperativity between plants and animals. Honeybees pollinate the flowers and keep the almond plants genetically diverse. Once pollinated, the flowers produce fruits and seeds; the seeds are the almonds we purchase.

# Chapter 19

# AP Biology Labs: Genetics

## Lab 7: Mitosis and Meiosis

The concepts of mitosis and meiosis are divided into a five-part lab; you may have done all parts or a few them. This section contains a brief review of each part.

### Part 1: Modeling Mitosis

Mitosis is cell division that is used for growth or repair. One cell divides into two genetically identical cells. For this activity, you used a material (such as clay, pop beads, or pipe cleaners) to represent chromosomes and moved the chromosomes through the steps of the cell cycle, mitosis, and cytokinesis. The main concepts you learned from this activity were:

- DNA duplicates in the S phase of interphase. After duplication, a cell contains double the amount of genetic material.

- DNA, in its chromatin form, condenses to make chromosomes in prophase. This allows for easier movement of the genetic material.

- Spindle fibers are used to separate sister chromatids. If sister chromatids fail to properly separate, the two daughter cells do not contain the proper number of chromosomes.

### Part 2: Effects of Environment on Mitosis

When you read about mitosis in a textbook or online, you are seeing the ideal situation for cell division. However, for an actual cell, the environment can significantly affect the rate of cell division. In Part 2, you investigated if a protein called lectin (a potential fungal pathogen) affected the number of cells undergoing mitosis in an onion plant. Lectins are known to increase mitosis in some root tips because rapid cell division weakens roots and harms agricultural crops; essentially, the presence of lectin in soil is problematic for the farming industry.

Your teacher provided you with untreated roots and roots treated with lectin. You squashed the root tips and stained the DNA. Then you counted the number of cells in mitosis and interphase and shared your data with the class. Class data was used for calculations to increase the reliability of the data and results. You may have calculated the mean and standard deviation for the data. Additionally, you calculated a chi-square value for your data. The chi-square formula is as follows:

$$X^2 = \frac{\sum(F_0 - F^e)^2}{F_e} = \frac{\sum(\text{Observed frequencies} - \text{Expected frequencies})^2}{\text{Expected frequencies}}$$

The chi-square table looked like:

|  | Interphase Cells | Mitosis Cells |
|---|---|---|
| # Observed |  |  |
| # Expected |  |  |
| $(F_0 - F^e)$ |  |  |
| $(F_0 - F^e)^2$ |  |  |
| $(F_0 - F^e)^2/F^e$ |  |  |
| Total: |  |  |

Because there were two groups (interphase and mitosis), there was one degree of freedom for your calculation (df = number of groups − 1). The probability value was 0.05, so with 1 df, the critical value was 3.84. If your calculated chi-square was greater than 3.84, you rejected your null hypothesis.

Exact calculations and numbers varied from lab to lab, thus cannot be given in this review. However, the number of cells in mitosis should have been higher in the lectin-treated root, while the number of cells in interphase should have been higher in the untreated root. From this data you inferred that the lectin-treated roots were dividing faster than the untreated roots.

## Part 3: Loss of Cell Cycle Control in Cancer

This activity involved comparing the karyotypes of normal cells with cancerous cells (specifically HeLa and leukemia cells). A **karyotype** is a picture of one cell's chromosome arranged by pair and number; it can show chromosomal abnormalities. HeLa cells are immortal cancer cells taken in 1951 from a cervical cancer patient named Henrietta Lacks. The HeLa cells show extra DNA due to the presence of DNA from a virus called human papillomavirus (HPV); this viral DNA is likely what caused the cancer. The leukemia karyotype shows a translocation, or a swap of pieces of chromosomes, between chromosomes 9 and 22. By comparing normal and cancerous karyotypes, it should be obvious that chromosomal abnormalities are found in cancer cells. In a normal cell, when DNA damage occurs, mitosis is stopped and the cell will not divide due to cell cycle control factors such as p53. However, cancerous cells often have damage to these factors and will divide despite the DNA damage, and, in fact, many cancers show that the p53 gene has been mutated so it cannot properly initiate cell death and stop cell division—hence, cancer cells' ability to grow quickly and uncontrollably.

## Part 4: Modeling Meiosis

This modeling activity is similar to Lab 1, except here you modeled meiosis. Like mitosis, meiosis involves duplication of DNA in the S phase of interphase. However, meiosis is cell division that results in gametes (sperm and egg) and it halves the chromosome number by starting with one diploid germ cell (2n) and ending with four haploid cells (n). There are several key factors about meiosis that increase genetic diversity in the gametes produced. For this activity, you used a material (such as clay, pop beads, or pipe cleaners) to represent chromosomes and moved the chromosomes through the steps of meiosis I and meiosis II. The main concepts you learned from this activity were:

- Chromosomes occur in homologous pairs that have the same genes, but not necessarily the same alleles. One homolog comes from the egg and the other from the sperm. An important difference between mitosis and meiosis is that the pairs synapse during prophase I. When the pairs synapse, **crossing over** can occur, in other words, exchanging genes between homologous pairs. Crossing over is one way meiosis increases genetic variation because it causes gametes to contain 4 different combinations of alleles.

- Meiosis I is the reduction division because this is where separation of homologous chromosomes occurs and is the step that leads to haploid cells. Meiosis II separates the identical sister chromatids.

- The other mechanism of meiosis that increases genetic variation is **independent assortment**. This means that homologous pairs separate randomly. There are $2^n$ (n = number of pairs) possible chromosomal combinations due to independent assortment. That means for humans, who have 23 pairs, there are $2^{23}$, or about 8 million, different gametes that can be made through independent assortment.

- If homologous pairs don't separate in meiosis I or sister chromatids don't separate in meiosis II, then the resulting gametes will have either an extra (triploidy) or a missing chromosome (monoploidy). If one of these abnormal gametes is used to create a zygote, the zygote will have an abnormal number of chromosomes. This is often not viable so the zygote will spontaneously die. There are a few exceptions, however, such as Down syndrome that results from a person who has three copies of chromosome 21.

## Part 5: Meiosis and Crossing Over in *Sordaria*

The rate of crossing over between homologous pairs is called the **recombinant frequency**. The farther apart two genes are, the more likely crossing over will happen. In this lab you used a fungus called *Sordaria* to calculate recombinant frequencies. You crossed a black and a tan strain of this fungus. When crossed, these fungi produce spore sacs, called asci, which contain 8 spores. Examining the spore patterns told you if recombination occurred or not. If there was no crossing over, the offspring had four black spores next to four tan spores. If crossing over occurred, the offspring had recombinant phenotypes that alternated colors in twos (2 black, 2 tan, 2 black, 2 tan), or 2 of one color, 4 of the other color, and 2 of the first color (2 black, 4 tan, 2 black).

After counting 50 asci, you calculated the distance between the color gene and the centromere in map units. One map unit equals one crossover event per 100 total events. Essentially, map units are used to tell how far apart genes are from one another. Your analysis table looked like this:

| # of Parental Asci | # of Crossover Asci | Total Asci | % Asci Showing Crossover | % Asci Showing Crossover/2 | Map Units |
|---|---|---|---|---|---|
| 24 | 26 | 50 | 52 | 26 | 26 |

Because the spores undergo mitosis (which is why there are eight spores, not four, following meiosis), you had to divide the crossover percentage by two.

# Lab 8: Bacterial Transformation

## Concepts

Although this lab used some complicated lab techniques, the science behind the activity is straightforward. Scientists have learned to manipulate organisms by removing DNA from one organism and transferring it to another. One example is the genetically modified food Bt corn. The Bt gene, which is found in a bacterium and produces a toxin that kills insects, was transferred to corn, giving corn resistance to insect pests.

In this lab, you transformed *E. coli* bacteria by inserting a plasmid. Transformation occurs when a cell takes up unpackaged DNA from the environment around it. A plasmid is a piece of extra-chromosomal DNA that is easily passed from one cell to another and can be taken up through transformation. The plasmid you used depends on your teacher's choice. However, the pGLO plasmid, described in the College Board lab manual, has three important genes on it. The first is the gene for green fluorescent protein (GFP) that glows when expressed. Expression of GFP is regulated by an inducible operon, so the plasmid contains a gene that turns on GFP production if the sugar arabinose is part of the nutrient agar. Finally, the plasmid contained a gene conferring ampicillin resistance.

## Science Practices

There are several inquiry-based approaches to this lab. This section, however, reviews only the basics of transformation. The first step was to use sterile technique (such as bleaching surfaces) to kill microorganisms that would contaminate your *E. coli* sample. To facilitate the uptake of the plasmid by *E. coli*, you used $CaCl_2$ and heat shocked the bacteria. Once you were done preparing the bacteria, you plated them on Petri dishes containing nutrient agar called LB (Luria broth). There were three types of agar used: plain LB agar, LB with ampicillin, and LB with ampicillin and arabinose. Your control plates contained non-transformed bacteria (−pGLO) while the transformation plates contained the transformed bacteria (+pGLO).

## Data Analysis

Ideal data would have been similar to that found in the table below:

| Plate Type | Observations |
|---|---|
| Control plate: LB  −pGLO | Many colonies or an even lawn of bacterial growth. Colonies/lawn are off-white in color. |
| Control plate: LB ampicillin  −pGLO | No bacterial growth. |
| Transformation plate: LB ampicillin  +pGLO | Many colonies of transformed bacteria. Colonies are off-white in color. |
| Transformation plate: LB ampicillin arabinose  +pGLO | Many colonies of transformed bacteria. Colonies are off-white in color, fluoresce green when exposed to UV light. |

Transformed cells were selected by growing the bacteria on agar containing ampicillin because any cell that successfully incorporated the plasmid would have the gene for ampicillin resistance. Cells that were grown on agar containing arabinose glowed green under fluorescent light because arabinose triggered the operon to express GFP.

After observing your transformed bacteria, you calculated the **transformation efficiency**, which tells you how effective you were in getting the DNA plasmid into the *E. coli* cells.

$$\text{Transformation Efficiency} = \frac{\text{number of transformed cells growing}}{\text{amount of DNA spread on the agar plate (in }\mu g)}$$

You counted the total number of green fluorescent colonies and recorded that as the total number of transformed cells. To determine the amount of DNA on the agar plate, you calculated the amount of pGLO DNA used in the experiment and the fraction of DNA you spread on the LB ampicillin arabinose plate.

## Conclusions

Scientists calculated the transformation efficiency of your pGLO lab between $8.0 \times 10^2$ μg and $7.0 \times 10^3$ μg. This number tells you the number of transformed bacterial colonies per microgram of DNA used. If your results fell between those numbers, you performed this lab well. If you numbers weren't similar, then you might have made a mistake in your lab procedures. Measuring transformation efficiency is important for patients undergoing gene therapy. Some cells are collected from a patient, transformed, and then put back into the patient. Scientists can use the transformation efficiency to determine how well the gene therapy will work for the patient.

## Lab 9: Restriction Enzyme Analysis

### Concepts

One of the most valuable advances in biotechnology is analysis of DNA through restriction enzyme digestion. This process creates a "DNA profile" that is used for paternity testing, identifying criminals and victims, and diagnosis of genetic disorders. There are two parts to this lab: digesting the DNA with restriction enzymes and separating the fragments of DNA by size through gel electrophoresis.

Restriction enzymes are isolated from bacteria and are used to make cuts in DNA. Each enzyme has a specific DNA sequence where it cuts usually a palindromic sequence (a nucleotide sequence that is the same if read from the same direction). For example, *Eco*RI makes a cut between the G and A in the sequence 5′-GAATTC-3′. See Figure 19.1 for an example of how this cut occurs.

Figure 19-1. How the Restriction Enzyme EcoRI Digests DNA

This cut leaves "sticky ends," which are places where nucleotides aren't paired. Because of this, scientists can cut DNA from one organism and place it into another organism if the same restriction enzyme is used because the sticky ends from one organism will match the sticky ends from the other organism.

Scientists can create DNA maps, called restriction maps, for an organism by digesting its DNA with restriction enzymes. Digestion will yield DNA fragments of various sizes; these fragments are called restriction fragment length polymorphisms (RFLPs). The RFLPs can be separated by gel electrophoresis, yielding a unique pattern. This can be used to compare DNA between individuals, compare different species' genomes, and determine evolutionary relationships.

## Science Practices

The DNA you used was probably already digested with restriction enzymes. Therefore, this section will focus on how you performed gel electrophoresis, which uses an electric current to separate charged particles by size. DNA is loaded into wells in an agarose gel, and an electric current is run through the gel. Because nucleic acids have a negative charge, they will migrate towards the opposite end of the gel, which has a positive charge. The smaller RFLPs will move faster because they can easily move through the porous gel, while larger fragments don't move far. Once the gel has run, it is stained so the DNA bands can be visualized.

## Data Analysis

Your gel probably resembled the photograph below in Figure 19-2. The larger shapes at the top represent large fragments that didn't move very far, and thus didn't separate well. However, the smaller fragments separate from one another and make clearer lines.

Figure 19-2.  Gel Electrophoresis

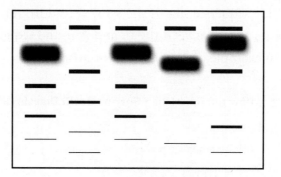

When we look at gels, we need to quantify our data by correlating how far a RFLP migrated with its size. Scientists often use base pair (bp) length to represent molecular weight; thus, a large bp number represents a fragment with a large molecular weight that didn't migrate far in the gel. To determine the bp lengths of unknown fragments, a standard DNA, with known bps, is run with the RFLPs and used to create a standard curve. You were given a photograph of an ideal gel run with a standard, and then graphed a standard curve. You placed distance traveled in centimeters on the $x$-axis and the base pair length on a logarithmic $y$-axis. Then you interpolated the length of the RFLPs using the distances traveled and your standard curve.

## Conclusions

If you were using your gel to match a criminal with a crime scene, then any DNA banding patterns that matched that of the crime scene would identify the criminal. For example, in Figure 19-3, you should compare the bands under "crime" to the bands for each of the three suspects to determine which one, if any, matches the bands from the crime. According to Figure 19-3, suspect 2 has bands that match the crime scene, therefore determining that suspect 2 in fact did the crime. This same principle is also used to identify relatives and even related species of organisms in evolutionary biology.

Figure 19-3. Banding Patterns Comparison

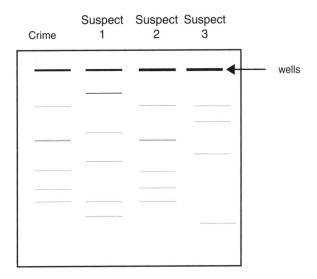

# Chapter 20

# Interactions

## Communities

Populations living within the same ecosystem interact with one another. The variety of species within the community depends upon the abiotic resources found in the ecosystem. Typically, a more diverse community having more variety and number of species is healthier than a community having only a few populations.

To better understand a community, scientists often use mathematical models to study the growth or decline of individual populations over time. A population that grows without limits is said to experience **exponential growth**, as seen in Figure 20-1. Exponential growth is characterized by slow growth during the initial time period, followed by rapid growth as time progresses. Most populations do not naturally exhibit exponential growth. Instead, as the **population density** increases, the number of organisms overwhelms the resources available in the ecosystem. Thus, some individuals will die due to starvation, thirst, disease, or predation. Eventually, population growth will level off at **carrying capacity**, the number of organisms that ecosystem can support. This type of growth model is called **logistic growth**. Carrying capacity is not a set value; it fluctuates with varying environmental conditions. If a population overshoots carrying capacity, its numbers will decrease. Conversely, a population beneath carrying capacity can grow.

Figure 20-1. Exponential and Logistic Growth Curves

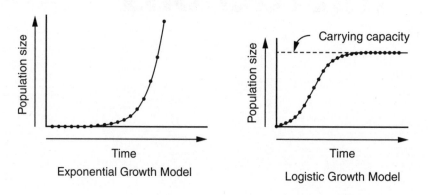

Factors that limit a population's growth are called **growth factors** and there are two types: density-dependent and density-independent factors. Density-dependent factors have a greater impact on a population with a higher density and include resources such as food and water. For example, a pond may contain enough water for 100 individuals; however, it cannot support a population of 200 individuals. Density-independent factors impact the population, regardless of the size. One example would be a frost that kills mosquitoes; it doesn't matter if there are 100 or 1,000 mosquitoes in the population because they will all be affected by the freeze.

### DIDYOUKNOW?

Humans are currently exhibiting exponential growth. This is worrisome to demographers (people who study human populations) because they are concerned that humans will outstrip Earth's resources and one day reach a carrying capacity.

### TEST TIP

You should be able to read population growth graphs and predict how changes to resources will affect the size of the population.

Another type of graph that is used to represent population growth patterns is called an **age structure diagram**. Demographers construct these using age distribution data and reproduction rates in order to study human populations. In Figure 20-2, the pyramidal shape of the graph on the left is representative of a human population that is growing. The number of individuals at younger ages indicates a population with many individuals in the reproductive age bracket who will have more children. The relatively flat shape of the graph on the right is representative of a population whose size is not fluctuating because all age groups are approximately the same size.

> **DID YOU KNOW?**
>
> There are a few countries in the world showing negative population growth. The shape of this age structure diagram is an inverted pyramid. Most of these countries are well-developed nations in Europe such as Italy and Russia.

Figure 20-2. Age Structure Diagrams for Two Human Populations

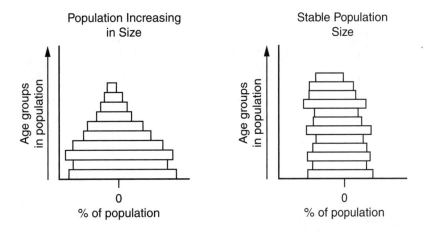

## Movement of Matter and Energy

Energy from the sun is the ultimate source of energy for terrestrial ecosystems. Energy flows into the ecosystem as light energy and is converted by **producers** into chemical energy. The amount of energy made by the producers in an ecosystem that is available to consumers is referred to as the **net primary productivity** (NPP) of the ecosystem. Environmental factors such as sunlight, rainfall, and temperature affect an ecosystem's NPP; for example, tropical rainforests have the highest NPP of a terrestrial ecosystem because they receive enough rain and sun under moderate temperature to generate sugars year-round. However, fluctuations in these abiotic factors due to factors such as global warming can change NPP and thus the composition of the ecosystem.

Chemical energy generated by producers is passed up a **food chain** to **consumers**. See Figure 20-3 for an example of a forest food chain. Each level of the food chain is called a **trophic level**. Note that decomposers act at all levels of a food chain. Some of the usable energy is converted into unusable energy, such as heat, thus the transfer of energy through an ecosystem is very inefficient. Approximately 10 percent of the energy available at one trophic level is available to the organisms at the next trophic level. That means that if 1000 kCal are found in the leaf matter of the forest, then only 100 kCal are available to the caterpillar that eats the leaves. Because of this "loss" of energy up a food chain, ecosystems cannot support many higher order consumers because there is not enough energy in the producer level to feed them. For this reason, diagrams of food chains are often shaped like a pyramid, with producers forming the base and upper level consumers at the top.

Figure 20-3.  Forest Food Chain

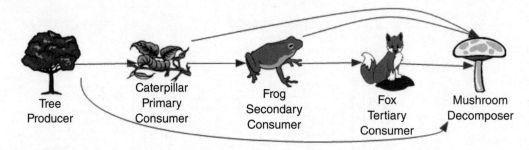

# TEST TIP

The AP Biology exam will have six "grid-in" questions that will be based on math problems. Calculating energy availability in a food chain or food web would be a great grid-in topic!

A food chain is a useful model for understanding the flow of energy in an ecosystem; however, it is simplistic and doesn't show all the relationships in an ecosystem. A **food web** shows multiple feeding relationships in an ecosystem. If snakes, hawks, squirrels, robins, deer, bacteria, and mountain lions were added to the food chain in Figure 20-3, it would be a food web. A food web is useful to predict how changes in one population's numbers might affect other populations. For example, it would help scientists understand how a dramatic decrease in the squirrel population might affect the population of hawks or mountain lions.

While energy flows through an ecosystem, matter must cycle because Earth is a closed system. Cells are mostly made up of water and organic compounds, thus water, carbon, nitrogen, and phosphorus cycles are important to ecosystem stability. Decomposers are the main component of all ecosystems for cycling matter. When organisms die, decomposers break down the organic molecules and return them to the soil or air for producers to absorb and integrate into their matter. Essentially, these elements cycle between organic molecules in living organisms and inorganic molecules in the abiotic factors of the ecosystem. Table 20-1 provides a quick summary of each of these cycles.

Table 20-1. The Four Major Energy Cycles

| Cycle | Key Features of the Cycle | Human Impact on the Cycle |
|---|---|---|
| **Water** — Atmosphere, Precipitation, Transpiration, Evaporation, Plants, Other organisms, Groundwater, Oceans, lakes, etc. | Major source is the ocean. Transpiration is evaporation in plants. | Pollution Overuse |
| **Carbon** — Atmosphere, $CO_2$, Photosynthesis, Cellular respiration, Combustion, Plants, Other organisms, Fossil fuels | Major source is the atmosphere ($CO_2$). Source of organic carbon. | Burning of fossil fuels moves $CO_2$ from ground to air. |
| **Nitrogen** — Atmosphere, $N_2$, Nitrogen fixation, Denitrification, Other organisms, Plants, Assimilation, Nitrogen-fixing bacteria, Ammonification, $NH_3$, $NO_3^-$, Nitrification, Denitrifying bacteria, Nitrifying bacteria | Major N source is the atmosphere ($N_2$). $N_2$ is not used by plants. Bacteria and legumes fix $N_2$ into usable forms of ammonia and nitrate. | Nitrogen use in fertilizer washes from soil into waterways, causing overgrowth of plants and algae. |
| **Phosphorus** — Other organisms, Plants, Weathering, Rock, Decomposition, Decomposers, P | Major P source is rock. The only cycle that does not have an atmospheric component. | Phosphorus use in fertilizer washes from soil into waterways, causing overgrowth of plants and algae. |

As the human population continues to increase exponentially, it is harming Earth's ecosystems so severely that the ecosystems may not recover. The current human population is approximately 7 billion, thus human demands on Earth are extraordinary. Humans need land, use natural resources, and pollute resources. Thus, they are causing massive decreases in populations of other organisms, often to the point of extinction. Because organisms in ecosystems are linked through their interactions, as we cause the decrease of one species, it disrupts the balance of the entire community, sometimes causing the collapse of the entire community.

> **DIDYOUKNOW?**
>
> Nitrogen- and phosphorus-based fertilizers used on farms in the Midwest wash into the Mississippi and into the Gulf of Mexico during the growing season each year. The resulting overgrowth of algae results in a "dead zone" in the Gulf that can stretch 6,000–7,000 miles. It is called a dead zone because when the algae die and decay, decomposers use all the oxygen in the water, causing larger organisms, such as fish and crustaceans, to die from oxygen starvation.

## Questions to Consider

1. A grassy area consists of grasshoppers, robins, mice, garter snakes, hawks and coyotes. Coyotes and hawks can eat robins, mice, and snakes. Snakes eat mice and robins' eggs. Robins and mice eat grasshoppers and grass seeds. Grasshoppers eat grass.

    Use the information above to construct a food web. If 10,000 kCal are available in the grass, how much energy would a coyote get if it acted as a tertiary consumer? A quaternary consumer?

    Finally, what important trophic level was left out of this scenario and why is it important?

2. Deer are considered to be pests by many people; they invade our backyards, eat our gardens, spread Lyme disease, and cause car accidents. Deer are prey for animals such as wolves, coyotes, mountain lions, and occasionally bears. Considering the growth of human populations over the past 200 years and the complexities of the food web, explain why deer have become such a problem.

## Answers:

1. The coyote would get 10 kcal of energy as a tertiary consumer and 1 kcal of energy as a quaternary consumer. The decomposers are left out of this scenario. They are important because they help cycle materials in an ecosystem.

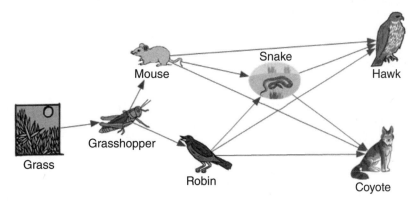

2. Deer predators are large animals that live at the top of the food web. They require large amounts of land in their territory in order to obtain enough energy to feed them. As humans have taken over more and more habitat, it has caused a large decrease in the number of these predators because they don't have enough space. Additionally, these predators are considered "dangerous" by humans and are often hunted and killed. The predators act to naturally control the deer population; by removing them from the ecosystem, humans have caused the explosion in deer population growth.

# Chapter 21

# Competition and Cooperation

## Molecular Interactions

Competition and cooperation from molecules to communities govern much of the biological world. For example, cooperation at the cellular level detects and responds to blood glucose levels. In turn, the response from one organ, the pancreas, is communicated to another organ, the liver, to further cooperate in the regulation of glucose levels. The cooperation and communication from molecules translates to the healthy activity of all cells that interact with glucose and ultimately to the health of the entire organism. At the molecular level, it is the *shape* of biologically active molecules that is key to the correct functioning of the molecule on the smaller scale and the entire organism on the larger scale.

Proteins carry out most of the molecular interactions within a living system, and, in each case, the interaction is based on conformational changes within the protein. Recall that proteins take on their working shape based on instructions from DNA as well as the environment in which they are formed and reside. The changes in conformation come about through interaction with other molecules.

An example of molecular interaction is illustrated by the action of proteins embedded within the cell membrane. One or more of the protein's **domains**—biologically active parts of the protein—may extend from the cell and act as a signal receptor. Its function as a receptor domain is dependent on having a sequence of amino acids that characterize it and are specifically complementary—either chemically or geometrically—to the signal it will respond to. However, the act of binding to the specific signal changes the conformation of the protein. The change in this exterior domain area causes a ripple effect change in the conformation of the rest of the protein, including the domain(s) that are embedded in the membrane and extending into the cytoplasm. The interior protein domain changes have the power to further interact with molecules within the cell, initiating changes within the cell. Variations in this sequence of signal reception, conformation change, and response is widespread and critical to most metabolic pathways.

Particularly important to carrying out metabolic functions within living systems are **enzymes**. Enzymes are protein catalysts—molecules that speed up the rate of a reaction by lowering the activation energy necessary to start a reaction. The graph in Figure 21-1 shows the difference in the activation energy of a catalyzed versus uncatalyzed reaction. Lowering the activation energy action enables chemical reactions to take place within conditions suitable for cells. Note that enzymes do not affect the overall energy changes of a reaction, just the activation energy.

Figure 21-1. The Energetics of a Catalyzed versus Uncatalyzed Reaction

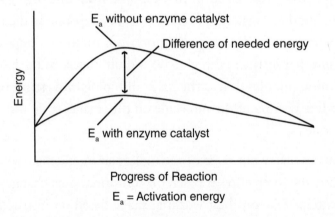

Enzymes are globular proteins at the tertiary or quaternary level of organization with one or more **active sites**. The active site is the small area on the enzyme that is complementary to, and, so, will bind to, the **substrate**, the reactants that the enzyme acts on. The complementarity of the active site of an enzyme to its substrate is determined by a few amino acids, so an enzyme is very specific and will catalyze only one reaction. The active site will not accept even an isomer of a molecule.

Reactants enter the active site of an enzyme through random collision events and, if it is the correct reactant, bind weakly with the active site to form the enzyme-substrate complex. Once in the active site, the reactants cause a conformational change in the enzyme that is called an **induced fit**. This puts reactants in a favorable orientation for bonding, stresses existing bonds, or places the reactants in a microenvironment—such as a lower pH—that favors the reaction. Once the reaction is complete, the products are released from the active site and the enzyme is free to catalyze another reaction. Figure 21-2 schematically illustrates the action of an enzyme.

> **DID YOU KNOW?**
>
> Ribulose bisphosphate carboxylase (Rubisco), the enzyme that catalyzes the first step in carbon dioxide fixation, is probably the most abundant protein in the natural world. It involves the interaction of the protein products of nuclear genes with the protein products of genes found in the chloroplast itself.

Figure 21-2. Schematic of an Enzyme Reaction Cycle

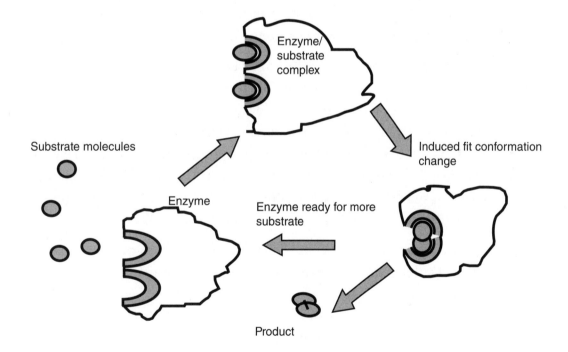

# TEST TIP

Questions about the specificity of protein shape and its effects on the action of the protein *will be* on the AP Biology exam. You should know how enzymes catalyze biological reactions and what environmental changes can alter the actions of enzymes. Even though memorization of particular enzymes is not necessary, really understanding one or two examples of metabolic pathways will help you generalize your understanding of enzymes and enable you to answer most questions about the importance of protein shape to their function.

Many enzymes require **cofactors** to carry out their tasks. Cofactors are not proteins, but they form critical associations with enzymes. The cofactor may be a metal ion or may be organic molecules called coenzymes. Some important coenzymes function in redox reactions. For example, $NAD^+$ and $NADP^+$ are coenzymes that act as electron acceptors in cell respiration and photosynthesis, respectively. These coenzymes are loosely bound to the enzymes with which they function. In glycolysis $NAD^+$ works with a dehydrogenase to receive two electrons and one hydrogen atom from the three-carbon compound that resulted from the first steps of glycolysis. Other coenzymes may act as carriers for functional groups or atoms.

**Enzymatic reactions** are affected by the conditions surrounding the reaction. Since substrates enter the active site randomly, the chances of a collision between an enzyme and its substrate are increased if the concentration of the substrate increases. In addition, many enzymes are arranged in a physical array that puts them into position to "meet" their substrate. Besides substrate availability, other environmental changes that can affect the rate of an enzymatic reaction include enzyme availability, temperature changes, pH changes, and inhibition or activation of the enzyme. In mammals, most enzymes work most efficiently at a pH of 6–8 and a temperature of 37 °C.

It is important to understand that the protein shape can be altered, temporarily or permanently, by conditions in the protein's environment. If the temperature of most enzymes goes above 40 °C, they will denature, that is, their structure will unravel. The same thing can happen with a change in pH. Some enzymes are actually produced in an inactive form that is activated by a change in the environment.

Inhibition of enzymatic action can be the result of a normal control mechanism for a metabolic pathway, or it can be an unnatural interference. Competitive inhibitors are molecules that vie with the normal substrate for the active site of the enzyme therefore blocking its entrance into the active site. This reduces the productivity of the enzyme. Some poisons are competitive inhibitors, and medicine has made use of competitive inhibitors to stop the action of bacterial enzymes disrupting their metabolic pathways. Noncompetitive inhibitors are molecules that bind to the enzyme at some site other than its active site, thus altering the protein's conformation, including the conformation at the active site. Again, enzyme productivity is reduced.

Most biological reactions take place in many steps, each step producing a product for the next step and each step catalyzed by a different enzyme. This sequential pathway provides an opportunity for a feedback mechanism that is used to control the overall metabolic pathway. Allosteric activation and inhibition are just such common feedback control mechanisms.

**Allosteric** control relies on the presence of a regulatory binding site on an enzyme that is separate from the active site(s). The **allosteric regulator** will occupy the regulatory binding site and change the conformation of the whole protein including the active site(s). If the regulator is an activator, the protein will be stabilized into an active form by the allosteric regulator. See Figure 21-3 for an example of allosteric activation. If the regulator is an inhibitor, its presence will change the conformation to an inactive form.

> **DID YOU KNOW?**
> Methanol poisoning occurs when methanol is oxidized to formaldehyde and formic acid, products which attack the optic nerve and cause blindness. Ethanol is an antidote because it is a competitive inhibitor of methanol, so ethanol slows the production of formaldehyde and formic acid.

Oftentimes, an allosteric inhibitor is the product of a metabolic pathway. This means that as the product is formed, some of it will occupy the allosteric regulatory site of an enzyme earlier in the pathway, inhibiting the enzyme pathway. As the product is used up, the molecules in the allosteric site will also be used and free up the enzyme to function again. This is an example of feedback inhibition, a common metabolic control.

Figure 21-3. Allosteric Activation

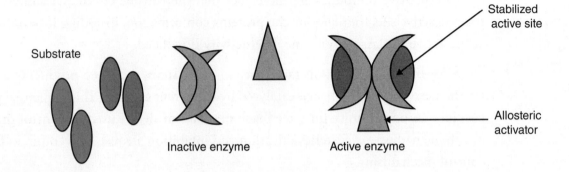

> **TEST TIP**
>
> Review the enzyme lab material about the effects of various environmental changes on enzymatically driven reactions. It is highly likely that there will be questions about this lab on the AP exam.

## Organism Interactions

Compartmentalization, within cells or organs, allows for multiple actions to occur simultaneously under differing conditions. On a larger scale, systems composed of organs provide compartmentalization for essentially the same reason that cells compartmentalize. Both at the cellular level and the organism level, there is constant activity that must be accomplished to support life.

Inside a cell, many activities are going on simultaneously to accomplish needed functions. Within a cell these specialized functions are carried out by the organelles. Table 21-1 summarizes some of the critical activities of life and which organelles cooperate to carry out each activity.

Table 21-1. Specialized Activities of Organelles

| Characteristic of Life | Example of Organelles Involved |
|---|---|
| Energy transformation | Mitochondria oxidize food molecules to produce ATP, a form of energy usable by biological molecules. Conversely, in plants chloroplasts convert light energy into chemical energy that can be processed by mitochondria. |
| Growth and development | mRNA, ribosomes, endoplasmic reticulum, Golgi apparatus, and cytoskeletal elements work to synthesize cellular components and products. |
| Heredity and reproduction | Chromosomes contained within a nucleus are copied and passed on to new cells in mitosis or to gametes in meiosis. |
| Homeostasis | The cell membrane separates the interior of the cell from its environment and permits selective traffic across the membrane. |

### TEST TIP

You should be able to match a cell type with some of its more important or predominant organelles. For example, you should know that a muscle cell will have a relatively large quantity of mitochondria because of the high energy requirement in muscle cells.

All living things must obtain and use energy. Plants produce their own usable energy through photosynthesis, but they must obtain certain nutrients from the soil. Many prokaryotes, some protists, fungi, and all animals must ingest food to obtain usable chemical energy and to supply essential nutrients for use as raw materials for their own maintenance and construction. In most animals, the digestive system extends from the mouth to the anus and is divided into compartments, each specialized to carry out a particular part of digestion. The digestive system will serve as an illustration of cooperation at the organism level. See Figure 21-4, the human digestive tract, as an illustration of the compartmentalization of the digestive system.

Figure 21-4. The Human Digestive Tract

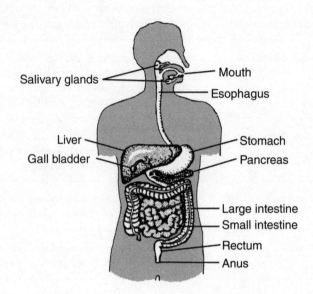

Among animals, digestion, the breakdown of food into its constituent molecules and the delivery of the molecules to the rest of the body, can be accomplished in two ways. Some animals, e.g., the sponges, take food into their body within a membrane-bound vacuole and the food is digested within this vacuole. This is known as *intracellular* digestion. Other animals digest their food outside of cells in *extracellular* digestion.

Digestion occurs in a sequence of steps in specialized compartments that proceed from the mouth to the anus. The food is moved along from the mouth through the esophagus to the stomach, the small intestine, the large intestine, the rectum, and out through the anus. The movement is carried by smooth muscle contractions called **peristalsis**. At points along the pathway, digestive compartments are separated by **sphincters**, ring-like muscles under the involuntary control of the nervous system. Sphincters control movement of the food material from one compartment to the next. Table 21-2 outlines food movement and digestive activities.

Table 21-2. Food Movement and Digestive Activities

| Component | Structure | Function |
|---|---|---|
| Mouth (oral cavity) | Teeth for tearing and grinding, tongue for tasting and manipulation | Ingestion; mechanical and chemical (starch) digestion (see salivary glands) |
| Salivary glands | Three pairs; mucus contains amylase | Starch digestion; moisten food |
| Esophagus<br><br>Pharynx/ epiglottis | Longitudinal and circular muscles for peristalsis; movable cartilage flap | Transports food from mouth to stomach; directs food to esophagus |
| Stomach | Three muscle layers; low PH, pepsin, protective mucus; cardiac and pyloric sphincter | Mechanical digestion; chemical digestion of protein; *bolus*[1] and *chyme*[2] movement |
| Pancreas (functions in digestion as an *exocrine gland*[3]) | Duct delivers enzymes and bicarbonate ions (low pH) to small intestine | Provides enzymes that digest proteins, lipids, and carbohydrates; raises pH of chyme |
| Gallbladder (not a gland) | Saclike; duct delivers bile to small intestine | Stores bile salts made by the liver that emulsify fats |
| Small intestine (sections: duodenum, jejunum, and ileum) | Walls of lining secrete mucus and enzymes; villi and microvilli increase absorptive surface area | Produces digestive enzymes for completion of digestion; absorption of nutrients and water |
| Liver | Large organ; has functions in different organ systems | Produces bile; stores energy as *glycogen*[4]; removes toxins |
| Large intestine (colon) | Ascending, transverse, descending (*cecum*[5] and *appendix*[6]); contains bacteria | Removes additional water; absorption of vitamins produced by bacteria |
| Rectum | Involuntary and voluntary sphincters near anus | Stores and eliminates solid waste (feces) |
| Anus | Posterior opening | Feces pass to exterior |

[1]*bolus*—ball of food that is swallowed; [2]*chyme*—name of partially digested material that leaves the stomach for the small intestine; [3]*exocrine gland*—gland that secretes substances via small tubes called ducts; [4]*glycogen*—storage polysaccharide of animals found in the liver; [5]*cecum*—a portion of the colon (generally smaller in carnivores and omnivores than in herbivores) that is involved in fermentation and processing plant material; [6]*appendix*—tiny fingerlike projection of the cecum in humans

Digestion is controlled by the actions of hormones. Gastrin is produced by stomach cells and acts on other stomach cells to increase production of digestive fluids. Enterogastrone is produced by the small intestine to slow digestion. Secretin and cholecystokinin are secreted by the small intestine and act on the pancreas and gall bladder to increase their secretions.

> **DID YOU KNOW?**
> If the lining of the small intestine were spread out, it would be approximately the area of a tennis court. An organism is a marvel of packing!

Since obtaining energy is vital to survival, animals have evolved to experience hunger. Like digestion, appetite is also under hormonal control. Leptin, ghrelin, and insulin are hormones that signal hunger or fullness.

The monomer units that are the result of digestion are now ready for use as building blocks of specific molecules or to use as a fuel in cellular respiration to generate ATP. Excess nutrients are converted into glycogen, which is stored in the liver, and fat, which is stored in specialized adipose tissue. Glycogen is easily converted to glucose in response to glucagon secreted by the pancreas. Fat stores are more slowly accessed.

Animals have evolved structures and processes that maximize their ingestion of food. Carnivores have teeth adapted for tearing, large stomachs adapted for large meals and long periods of time between meals, and a relatively short small intestine since protein digestion is simpler than vegetable material digestion.

Cellulose is the main molecule of plant cell walls. It is a starch that, due to its structure, is undigestable by most animals—they do not possess the enzymes necessary to hydrolyze cellulose. Many animals have evolved mutualistic relationships with microorganisms that do have enzymes that can break down cellulose. Cows, horses, and rabbits (and termites) are examples of common organisms that rely on bacteria that reside in specialized chambers within their digestive systems to hydrolyze cellulose. The animal obtains energy from an otherwise unusable source and are often supplied with vitamins not present in their diet. In return, the bacteria is housed and continuously supplied with an energy source.

Interactions among populations of microorganisms can result in a community of microorganisms that act as if they are a single, multicellular organism. A biofilm, a layer of microorganisms attached to a surface and covered by a coating, is an example of a community of microorganisms. The interactions among the members of the biofilm increase their metabolic efficiencies and resistance to environmental stresses or inhibitors, thus boosting their survival. Chronic wounds illustrate this community interaction.

Wounds are initially populated by opportunistic microbes suited to the wound site. These colonizers modify the environment making it suitable for other kinds of microbes. For example, obligate anaerobes can develop in a wound even if it is frequently exposed to air because of the presence of aerobic bacteria consuming oxygen. As the interactions continue and more microbes are established, a patchwork of microenvironments allows the survival of many different component species. A simple food chain may develop in which the product of one group of microbes becomes the energy source for another group. Growth factors released by one group may influence the cell division activity of all the resident groups of microorganisms. The different members of the community may provide enhanced survival functions for all of the members by such activities as the spread of antibiotic resistance and the increased production of toxins and destructive enzymes. Ultimately, an interdependent community is established, and, in the case of wounds, it makes treatment very difficult.

> **TEST TIP**
>
> It is not necessary to memorize any of the enzymes described or even the digestive system. No one system or set of enzymes will be on the AP Biology exam. The enzymes are mentioned for the clarity of the explanation and because it illustrates the integration and coordination that goes on at each level of biological organization within an organism. It would be smart to recall that the names of most enzymes end in *–ase*.

## Population Interactions

Populations don't act alone; instead they interact to form a functional **community**, the biotic portions of an ecosystem. The health and stability of the community depends upon the way each population in a community functions. Additionally, the interactions between populations affect the distribution and size of each of the individual species within that community.

**Symbiotic relationships** occur between two different species that live closely together. Examples are **parasitism**, in which the host organism is harmed, **mutualism**, in which both organisms benefit, and **commensalism**, in which one organism benefits and the other is neither helped nor harmed.

**Predation** occurs when a **predator** hunts and kills another species, the **prey**. Although it may seem like the predator would drive this interaction, mathematical models of complex predator-prey relationships over time indicate that the size of one population drives the size of the other population. For example, a crash in the prey population size will cause a crash in the predator population because each population is dependent upon the other. The figure below provides an example of deer and wolf populations; the wolf population numbers lag slightly behind any changes to the deer population.

Figure 21-5. Predation

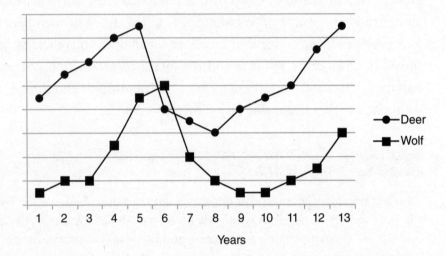

**Explanation of Figure 21-5:** Essentially, there is a feedback mechanism governing the size of each population. This feedback maintains stability in the ecosystem because it prevents one population from becoming too large and utilizing too many resources. For example, if the wolves were removed from the ecosystem, it could cause a population explosion in the deer. The deer could then over-consume their resources, leading to the starvation of many deer.

**Competition** between two different species for resources is called interspecific competition. Organisms may compete for water, sunlight, and food. The competitive exclusion principle states that two species cannot compete for every resource because one species will out-compete the other and come to dominate the ecosystem. Competition can also be modeled mathematically and shows a similar pattern to that of predatory-prey relationships.

The overall character of a population is not the same as the individuals comprising the population. A population may be well suited to its habitat and niche, even though some individual members have alleles that are less well suited. Cooperation and competition between individuals builds the character of the population. For example, the individuals with less suited alleles may be out-competed by members of their own species or killed by predators, resulting in a loss of this allele from the population over time. Essentially, these factors drive the evolution of individual populations to make them better suited to their environment.

# TEST TIP

The AP Biology exam focuses on synthesis of information from different content areas. Note that evolution and ecology are closely tied together, so you can expect some questions that will link concepts from both areas.

In addition to interactions between species, there are many external factors that influence the distribution of populations within an ecosystem. Some of these factors are:

- Decrease or increase in an abiotic resource—If there is a drought, plant species with shallow root systems may die out. However, an excess of rain might cause an increase in their population size.

- Environmental catastrophes—A flood might wash away all plant species having shallow root systems.

- Increased human activities—Deforestation of rainforests in order to grow crops, such as sugar cane, changes the distribution of species. Tree-dwelling species like birds would not be found in the ecosystem; however, the number of ground-dwellings organisms, such as amphibians, would increase.

- Invasive species—Non-native organisms often take over an ecosystem and decrease the number of native species. One example is kudzu, a leafy vine native to Asia. It has killed many tree species in the southeastern U.S. because it drapes over them, preventing their leaves from obtaining the sunlight necessary to photosynthesize.

## Distribution of Ecosystems

Just as population size and distribution is not constant, ecosystem distribution changes over time and is especially susceptible to human impact, geological events, and meteorological events. These changes can be small scale and spread slowly over hundreds or thousands of years, or they can be larger, as represented by the meteor impact that is believed to have contributed to the extinction of dinosaurs.

Currently, the biggest cause of changes to ecosystem distribution is human disturbance. Humans are burning and clear-cutting large swaths of tropical rainforest for cropland. In doing so, they are destroying a fragile ecosystem that has evolved for many thousands of years and that will not be able to recover. There are few nutrients in rainforest soils; instead, the nutrients are found in the trees. By burning the forests, the nutrients are lost and the soils' nutrient levels are exhausted after only a few years of farming. Loss of plant cover then leads to large scale erosion of soil, which washes into nearby rivers, polluting them. Thus, what seems like a simple action changes both the forest and the river ecosystems.

Another example is the growing desertification of lands around the Sahara desert in Africa. Due to overgrazing, marginal grasslands are being converted into desert landscape. As with the rainforests, this is a change that is likely permanent. One final example of human impact is caused by global climate change. As the planet gradually warms and rainfall patterns change, ecosystems that thrived in areas for thousands of years will shift to new geographic locations.

Geological events are usually studied through biogeographical data, such as where specific living organisms are distributed around the world. An important example of this is continental drift and how it has impacted the history and locations of certain animals around the world. For example, plant fossils of identical species have been found on the coasts of South America and Africa when these continents were connected as one land mass. These fossil remains provide clues to the distribution of different ecosystems throughout the Earth's history.

## Questions to Consider

1. Sketch a hypothetical, enzymatically driven pathway that requires 3 enzymes and is under feedback inhibition control at enzyme 1 by the product of the pathway. Explain your sketch. Is this an example of positive or negative feedback?

2. What happens to the enzyme pepsin, secreted in the stomach, when it reaches the small intestine? What is unusual about the working conditions for the enzyme pepsin?

3. Isle Royale, Michigan, is an island with a unique predator-prey dynamic consisting of a population of moose and a population of wolves. The wolves eat only the moose because there are no other large animals on the island. Most of the moose's

diet consists of a tree called the balsam fir. If humans decided to build a resort on Isle Royal and remove one-third of the island's fir forests, explain the impact on both wolves and moose.

4. Marsupials, such as kangaroos, opossums, and koalas, are found on the continents of Australia and South and North America. Additionally, their fossils have been found on Antarctica. Kangaroos and koalas on Australia live in a dry ecosystem, while the opossums in the Americas live in forest ecosystems. These animals are not strong swimmers, and there is no migration route between these continents. Explain how the distribution of this subset of mammals is indicative of changes to ecosystems over time.

## Answers:

1.

Reactants → ~~Enzyme A~~ → Intermediate 1 → Enzyme B → Intermediate 2 → Enzyme B → Product (with feedback arrow from Product to Enzyme A)

The diagram shows the product of this pathway inhibiting the action of Enzyme A. The product is probably an allosteric inhibitor of Enzyme A. This mechanism allows for negative feedback and prevents production of too much product, which wastes the resources of the cell or organism.

2. The digestive tract is separated into compartments. Each compartment operates under different conditions. The stomach, where the pepsin is secreted and functions, has a pH of 2. It is unusual for a protein to have such a low optimal pH. When the pepsin enters the relatively alkaline small intestine, the pepsin conformation is disrupted, the enzyme is denatured and no longer works to hydrolyze proteins.

> **DID YOU KNOW?**
> Isle Royale is the largest island in Lake Superior. It is actually a protected U.S. National Park, as well as a recognized Wilderness Area. Development as mentioned in Question 3 would not happen under current protective measures.

3. Because the primary food source of the moose is removed, many moose will be weakened due to starvation. Because of the weaker moose, the wolf population might increase due to the ease of prey capture. However, as many moose die of starvation or are eaten by the wolves, their population numbers will drop. This drop in population will cause the wolf population to decrease as well. At a certain point (carrying capacity), the moose population is likely to stabilize at a smaller number based upon the availability of food left on the island. As it stabilizes, the wolf population size should also stabilize.

4. About 160 million years ago, these continents were joined together and the southern portion of South America was at a similar latitude to Australia and Antarctica. Thus, marsupials radiated out from one of these areas. As the continents separated, climate conditions changed and the marsupials ended up located in and adapting to very different ecosystems, from the tropical rainforests of South American to the dry chaparral of Australia.

# Chapter 22

# Environmental Interactions

## Molecular Variation

The instructions on DNA are responsible for the tremendous variety of molecular compounds that are produced in an organism. Within any class of molecules, there are variations that confer different characteristics such that their function is further broadened. Furthermore, interactions between molecular variations and the environment alter the molecules and provide cells and organisms with an even wider range of functions.

An example of variations within a class of molecules is the difference between two lipids—a triglyceride and a phospholipid. A **triglyceride** is a glycerol molecule bound to three fatty acids. This structure makes the triglyceride very insoluble in water. However, exchange a negative phosphate group for one fatty acid chain and the molecule becomes a phospholipid. A **phospholipid** molecule exhibits polar and nonpolar characteristics that make it a suitable molecule for formation of the cell membranes. Further differences in the remaining two fatty acid chains can alter the resulting phospholipid and, consequently, the characteristics of the cell membrane. If the remaining fatty acid tails are unsaturated, kinks form in the fatty acid tails that prevent the phospholipids from packing as tightly as would saturated fatty acid tails. This lower packing density helps maintain cell membrane fluidity at lower temperatures.

One group of molecules that exhibits incredible diversity is the **antibody**. An antibody is the protein product of a B cell, one of a group of cells called **lymphocytes**. The human immune system has the potential to make over a billion different antibodies, yet the human genome has only about 20,500 genes. The origin of the differences in antibodies is the genetic rearrangement that occurs during B cell maturation and differentiation. A B cell is a lymphocyte that is characterized by a specific antigen receptor. Once a B cell is activated by the presence of an antigen, it produces antibodies. Antibodies are proteins that are like the receptor proteins found on the B cells from which they are produced, except that instead of the chain that anchors the receptor in the cell membrane, the antibody has a region called a constant region. The constant region determines where and how the particular antibody will act. There are five classes of antibodies, each acting in a different manner to neutralize a pathogen.

An antibody consists of two identical light chains and two identical heavy chains. Both the light and heavy chains have a variable region, a joining region, and a constant region.

Each light chain can have one of 40 (or 30 dependent on the kind of light chain) different variable regions coded for by 40 different DNA segments. There are also five different joining segments. There is only one constant region, but which one it is is dependent on the class of **immunoglobulin** (the antibody) that the cell can form. During B cell development, **recombinases**, enzymes that cut and rearrange DNA, randomly link a variable region to a joining region by removing the DNA between the two regions. Once these regions have been linked, the B cell will be committed to producing the light chain of the antibodies with this particular combination of variable and joining regions. If this committed B cell is activated, the variable and joining regions will be transcribed along with the constant region that is adjacent to the joining region on the DNA. After transcription, the intron between the joining and constant regions will be removed. The light chain transcript will then be processed as a secretory protein. Variability in antibody receptor sites is further increased because within the same B cell as the light chain, a heavy chain goes through the same combinatorial joining as the light chain. But, a heavy chain has even more genetic variations and more parts to come together. The variations are further increased because of mutations that can occur.

Other events can occur within the genome that can contribute to variations in phenotypes. Such variations include having multiple alleles or duplicate copies of genes. For example, hemoglobin is the oxygen-carrying molecule of the blood. It consists of four polypeptide chains—2 alpha and 2 beta chains. Fetal hemoglobin also has four subunits. They are 2 alpha and 2 *gamma* subunits. The gene for the gamma subunit is on the same chromosome as the gene for the beta subunit and differs only slightly. This difference gives fetal hemoglobin a higher affinity for oxygen than does adult hemoglobin. This is a survival advantage because it increases the efficiency with which a fetus can extract oxygen from the mother's blood. The hemoglobin family probably arose by duplication and transposition events.

One hypothesis about duplicated genes is that once a gene has been duplicated, the duplicate can respond to various selective pressures separately from the original gene and proceed along its own evolutionary pathway. The Antarctic eelpout, *Pachycara brachycephalum*, is a cold-blooded organism that produces a kind of antifreeze for its blood. Studies have shown that the antifreeze protein is almost identical to sialic acid synthase B (SASB), a protein that has the capacity to prevent ice formation. However, SASB does not leave the cell to circulate in the blood. Small genetic manipulations of SASB by researchers enabled the molecule to become a secretory protein and circulate within the blood. This is evidence that a gene duplication took place within the eelpout with the duplicate gene following its own evolutionary pathway. Changes to the duplicated gene enabled it to become a secreted protein conferring antifreeze properties to the circulating blood.

**DID YOU KNOW?**

The specificity of antigen-antibody interactions are exploited in commercial ways. The exploitation relies on using designed antibodies to "tag" specific molecules. These types of assays can be used in detection of conditions that produce specific molecules. The most common pregnancy test, for example, relies on **monoclonal** antibodies that bind to and tag human chorionic gonadotropin (HCG), a chemical produced as soon as an embryo implants in the uterus.

**Heterozygosity**, having two different alleles present for a particular trait, increases phenotypic possibilities. For example, one amino acid base change in the primary structure of hemoglobin can produce a faulty hemoglobin molecule and, if the individual is homozygous for the mutations that causes this change, the individual would have sickle cell anemia. Nature usually weeds out variants that do not provide survival benefits. Yet, in areas of the world where malaria is prevalent, the gene for sickle cell anemia has a relatively high frequency. This is because this heterozygote condition is a phenotypic variation with increased survival advantage over the homozygous dominant individual. If heterozygosity is advantageous, an otherwise deleterious allele will be maintained within the gene pool of the population.

**TEST TIP**

You will need to know that the great diversity of molecules—especially proteins—results from more than the action of the genetic code. Alternative RNA splicing, for example, enables one transcript to code for more than one kind of protein.

# Environmental Effects on Genotype

Variation in traits is not always due to variations in genes. Genes dictate the range of phenotypes possible for an organism. But within the range, the environment plays a role in how the phenotype is ultimately expressed. This phenotypic plasticity in response to the environment may provide variations that promote survival.

In some arctic animals such as the arctic fox, a gene is present that produces an enzyme to catalyze production of coat color. However, it is an enzyme that is **thermolabile**, that is, unable to work at even a slightly higher temperature than its ideal temperature. An arctic fox is brown in the summer and white in the winter. In the warmth of the summer, the coat developing is white because the thermolabile enzyme is not functioning. In the winter, the coat developing is brown because of the now functioning enzyme. By the time the seasons change, the coat color is established—white in winter and brown in summer. Another example of thermolability occurs in the Himalayan rabbit, which has a gene for producing coat color that is inactive above 35 °C and is maximally active between 15 °C and 25 °C. If the rabbit is raised at 30 °C, it will be all white. At a lower temperature, its extremities will be black.

Other environmental changes, such as day length or seasonal changes, can also change the range of the phenotypic expression of identical genes. Yarrow plants that are cloned—thus genetically identical—will show different phenotypes when grown at different altitudes. Hydrangeas show a range of colors from pink in acidic soil to blue in basic soil. The broadest range of phenotypic variations usually occur with polygenic characteristics such as height in humans. Long-term studies of identical twins and identical armadillo quadruplets—armadillos are always born as four identical pups—show that these identical organisms exhibit very different phenotypes and that phenotypic differences accumulate over time. These kinds of changes are thought to be due to epigenetics, reversible heritable changes that are not due to a change in the DNA. Epigenetics is thought to be a link between the environment and gene expression. An example of epigenetic action is the silencing of a gene by methylation. Methylation and the silencing of genes occur naturally as part of cell differentiation processes, but it has been shown to occur in response to environmental changes.

## Population Dynamics

Genetic diversity is the variety of allele types found within a population. Examples of populations whose species may have little genetic diversity include asexually reproducing organisms or populations having small sizes. These populations having little genetic diversity are at risk of extinction. One example of a population with little genetic diversity is the cheetahs living in Africa. Around 10,000 years ago, it appears there was a population bottleneck that drastically reduced the size of the cheetah population. Additionally, in modern times, they must compete with humans for territory. This has left the current cheetah population, numbering around 12,000 individuals, with little genetic diversity and susceptible to disease.

> **DIDYOUKNOW?**
> Some cheetahs are so genetically similar that a skin graft from one animal can be placed onto another animal with little to no immune response, indicating incredibly high similarities in their immune systems.

The importance of genetic diversity is that it allows for differential responses to environmental changes. If there is an environmental change, the population must have sufficient diversity for some organisms to survive the change. Returning to the cheetah example above, if most cheetahs have similar immune systems, any pathogen that is capable of overcoming one cheetah's immune system will likely affect all the cheetahs, potentially devastating or exterminating the population. Instead, if a population has sufficient diversity, then an epidemic within the population would likely kill those organisms with non-resistant genes, sicken some with slightly resistant genes, and not affect those with highly resistant genes. Thus, although the population numbers might decline, the population will not be forced to extinction.

One way to measure genetic variation in a population is through the use of the Hardy-Weinberg equation. The equation measures the frequencies of the dominant allele of a gene ($p$) and the recessive allele of a gene ($q$). Essentially, it can be used to model variations in alleles in a population. In a population not undergoing evolution, the genetic variation ($p$ and $q$) should remain constant over time. A population that shows fairly similar values of $p$ and $q$ is more stable than a population having an extreme value for $p$ and $q$ because in the extreme example, one allele could easily be eliminated through a change in the environment (**genetic drift**).

> **TEST TIP**
>
> There is no reason to memorize the Hardy-Weinberg equation because it will be provided for you on the formula sheet. However, it is an important concept for both evolutionary biology and population genetics, so you should be able to solve problems using the equation and understand what each of the components means.

# Biodiversity

Diversity of species within an ecosystem is an important ecosystem stabilizer. The variety of species within an ecosystem is known as **species biodiversity**. Typically, an ecosystem that has many species is more stable than one having a few types of organisms. This is because individual species provide many services to ecosystems. For example, insects pollinate plants (and our crops); worms, bacteria, and fungi decompose dead organic matter; and plants and algae convert carbon dioxide into oxygen. Taken together, all the organisms comprising an ecosystem function as a system and removing a species from this system makes the ecosystem more vulnerable to disturbances such as fire, disease, or human-caused changes. Diversity of ecosystems is maintained by several factors including:

- **Abiotic factors** are the nonliving physical and chemical components of an ecosystem, such as air, temperature, water, and soil nutrients. Specific ecosystems have a set requirement of these factors; for example, a rainforest receives a certain amount of rainfall a year. Any changes to the abiotic factors can threaten the survival of the species that are dependent upon these factors.

- **Biotic factors** are the living components of an ecosystem, such as plants, fungi, and decomposers. The living organisms are dependent upon one other in order to survive. **Producers** are the biotic factors that determine ecosystem type. For example, trees make an ecosystem a forest and provide food for some organisms, habitat for others, and oxygen for all. Producers form the base of the food chain and all organisms are ultimately dependent upon them as the source of food energy in the food chain. Thus, removal of the dominant producers from an ecosystem will cause it to collapse. This is one of the reasons why human destruction of the tropical rainforests is rapidly decreasing biodiversity in the tropics.

- **Keystone species** are species that have an important role in their ecosystem. They are usually not the most numerous species, but their niche is so significant that their removal causes damage to, or even collapse of, the ecosystem. One example is sea stars that feed on mussels in intertidal coastal regions. Removal of the sea stars allows the mussels to dominate the intertidal rocks and eliminate most other algal and invertebrate species, causing a drastic decrease in biodiversity. Essentially, an ecosystem that has its keystone species present shows more biodiversity than an ecosystem that is missing its keystone species.

> **DID YOU KNOW?**
> Keystone species are named for the keystone in an arch. This is the wedge-shaped piece of rock found at the top of an arch. It functions by holding the other stones together and makes the arch a weight-bearing architectural structure. Similarly, a keystone species acts to hold an ecosystem together.

Figure 22-1. Biodiversity

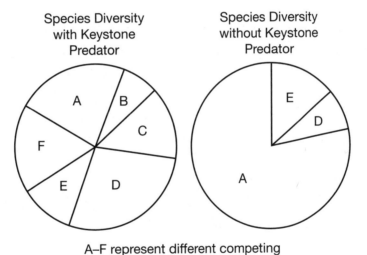

A–F represent different competing species in a community

## Questions to Consider

1. How do you think the cell membranes of cold-blooded fish living in the cold, deep Antarctic Ocean differs from the cell membranes of humans?

2. Two populations of mice in neighboring fields are being studied. The relative frequency of light alleles ($p$) is being compared to that of dark alleles ($q$), which is recessive to the light color. Using the data provided in the table, explain which population shows greater genetic diversity and which population is more stable.

|  | Population – Field 1 | Population – Field 2 |
|---|---|---|
| Frequency of $p$ | 0.60 | 0.95 |
| Frequency of $q$ | 0.40 | 0.15 |

3. Orcas (killer whales) on the Pacific coast will eat sea otters if their normal prey is not available. Sea otters eat sea urchins and the urchins eat kelp. Kelp forms underwater forests, which like the tropical rainforests are home to a wide diversity of organisms. In the 1980s and 1990s, a measured decline in the orca's normal food ultimately caused a significant decrease in the biodiversity of the ecosystem. Explain this phenomenon.

## Answers:

1. Both organisms will have cell membranes of phospholipids; however, because of both the cold and pressure, and the fact that the fish are cold-blooded and cannot regulate their temperature, the fish will be in danger of their cell membranes solidifying. To counter this, fish have largely unsaturated fatty acid tails in their cell membranes that maintains fluidity. Humans, being warm-blooded, do not subject their cell membranes to such extremes and have more saturated fatty acid tails in their membranes.

2. The population in Field 1 is more diverse and stable because it has a more even distribution of alleles. The population in Field 2 is more likely to experience a chance event (genetic drift) that could remove the $q$ allele from the population, thus destroying genetic diversity.

3. Sea otters are the keystone species in this ecosystem. They control the size of the sea urchin population, thus when their numbers dropped due to orca predation, the sea urchins increased and decimated the kelp. Loss of the kelp meant that the organisms which rely on the kelp for survival died, thus causing the decrease in biodiversity.

# Chapter 23

# AP Biology Labs: Biological Interactions

## Lab 10: Energy Dynamics

> **TEST TIP**
>
> There is a good chance you didn't do this lab because it involves a lot of set-up, and the maintenance of living organisms is complicated. However, it is important that you review the content of this lab because energy transfer is an important concept that is sure to appear on the exam.

## Concepts

All living things require energy in order to carry out their cellular functions, and the sun is the ultimate source of energy for the majority of life on Earth. **Producers**, in this lab Wisconsin Fast Plants®, convert sunlight energy through the process of photosynthesis into chemical energy and **biomass**. Biomass is the mass of organic matter in a living organism and is measured by drying the organisms in order to remove the water. **Gross primary productivity** (GPP) is a mathematical measure of the amount of light energy that is converted into chemical energy (in biomass) for these plants. However, some of the light energy the plant receives and converts into chemical energy is used in cellular respiration, some is needed for growth and maintenance, and some is converted into waste products and "lost" from the system, such as heat energy. Thus, **net primary productivity** (NPP) measures the amount of energy stored in plant biomass that is available to consumers in the ecosystem that consume the plants. Mathematically, NPP = GPP − respiration. Refer to Figure 23-1 to see the flow of energy for producers.

Figure 23-1. The Flow of Energy Into and Out of a Plant

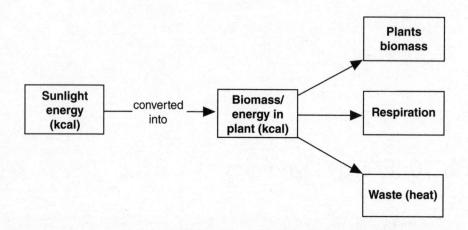

**Consumers** cannot produce their own energy and thus gain energy by eating producers, so the biomass generated by the plants (calculated from NPP) is important for ecosystem maintenance. In this lab the consumers were cabbage butterfly larvae. Although they did not eat the fast plants, they ate Brussels sprouts, which are in the same cabbage family as fast plants. Thus, we can draw a **food chain** to represent the flow of energy into the system representing an ecosystem in this lab. This simple food chain would resemble Figure 23-2.

Figure 23-2. Food Chain for Energy Dynamics Lab

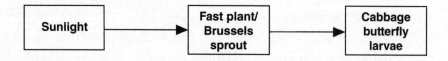

## Science Practices

There are few methods from this lab that you should know because maintenance of fast plant and butterfly cultures is not necessary for the exam. However, you should understand the fast plants were dried to remove all water in order to calculate biomass. Additionally, the larval frass (insect feces) was collected and measured because this waste product resulted from cellular respiration and had to be considered to accurately measure energy transformation.

## Data Analysis

The first part of the lab used fast plants. Your lab manual gave explicit directions about the data to collect and how to analyze it through calculations. Your data and analysis should look similar to the table below.

| Plant age | Mass of 10 plants | Dry mass of 10 plants | Percent biomass (dry/wet) | Energy in 10 plants (biomass x 4.35 kcal) | NPP per day per plant |
|---|---|---|---|---|---|
| 7 days | 20.1 grams | 4.4 grams | 21.9% | 19.4 kcal | 0.28 kcal/day |
| 14 days | 38.9 grams | 9.4 grams | 24.1% | 40.89 kcal | 0.29 kcal/day |

From this data, each of your plant's NPP was approximately 0.3 kcal each day. In a functioning ecosystem, these calories would be available to the consumers.

The second part of the lab used 10 cabbage butterfly larvae eating Brussels sprouts. As with the fast plants, your lab manual told you exactly what to collect and how to do the calculations required for analysis, thus refer to it with questions about the numbers found in the table. For example, steps 1-3 were calculated using your data from the fast plants, and the manual stated biomass of larvae is calculated as 40% of the wet mass of the larvae. Your data and analysis should look similar to the table below.

| Larvae age | 12 days | 15 days | Over 3 days of growth |
|---|---|---|---|
| 1. Mass of Brussels sprouts | 31 g | 11 g | 20 g consumed |
| 2. Dry mass of Brussels sprouts (mass × 0.23 – average % from fast plant data table) | 7.13 g | 2.53 g | 4.6 g |
| 3. Plant energy (biomass × 4.35 kcal/g) | 31.01 kcal | 11.00 kcal | 20.01 kcal |
| 4. Plant energy consumed by each larva (plant energy/10) | 3.1 kcal | 1.1 kcal | 2.0 kcal per larva |
| 5. Mass of 10 larvae | 0.4 g | 2.0 g | 1.6 g gained |

*(Continued)*

(*Continued*)

| Larvae age | 12 days | 15 days | Over 3 days of growth |
|---|---|---|---|
| 6. Average mass of one larva | 0.04 g | 0.2 g | 0.16 g gained |
| 7. Larvae percent biomass as stated in lab manual | 40% | 40% | 40% |
| 8. Individual larva biomass (mass × % biomass) | 0.016 g | 0.08 g | 0.064 g gained |
| 9. Energy produced per individual (individual biomass × 5.5 kcal/g) | 0.088 kcal | 0.44 kcal | 0.35 kcal |
| 10. Mass of the frass from all larvae | | 0.4 g | 0.4 g excreted |
| 11. Frass per individual | | 0.04 g | 0.04 g excreted |
| 12. Waste energy (frass mass v 4.76 kcal/g) | | 0.19 kcal | 0.19 kcal excreted |
| 13. Estimate of cellular respiration (plant energy consumed − waste energy produced) | | | 1.81 kcal |

From the data table, you can see that each larva consumed 2.0 kcal of energy from the plants, of which 1.81 kcal of energy was used in cellular respiration, some of which was converted into the 0.16 grams of mass gained by each larva.

The final part of this lab involved designing your own experiment. If you conducted this, then you asked a question such as "Do all plants have the same percent of biomass?" or "How much biomass do plants expend on making flowers?" You analyzed your data in a manner similar to the fast plants.

## Conclusions

In this lab you conducted an energy audit by considering the amount of energy going into a system and the amount of energy coming out, in all its various forms. For the plants, it was not possible to calculate the energy going in because that was light. However, the energy going into the larvae was possible to estimate by determining the amount of Brussels sprout biomass consumed by the larvae. In both plants and larvae, it was not possible to estimate the amount of energy given off as heat. However, you measured every aspect of energy transformation that was possible, so you should understand that light energy is converted by producers into chemical energy in photosynthesis. That chemical energy is then used by both producers and consumers in cellular respiration to do cellular work, to add biomass to organisms, and to produce waste (frass in larvae).

# Lab 11: Transpiration

## Concepts

Water is one of the required elements for photosynthesis, and plants obtain this water through their environment. The roots of plants absorb water, minerals, and ions from the soil, and these materials are transported through the vascular tissues, specifically the **xylem**, of the plants. Not only do plants uptake water through their roots, but they also lose water through evaporation from their leaves. The leaves of vascular plants have small openings called **stomata** that allow for gas exchange. Stomata open in order for carbon dioxide to enter the leaf and oxygen to leave; however, when they are open, it allows the evaporation of water through the leaf in a process called **transpiration**. A plant must maintain homeostasis by balancing its needs for gas exchange with its loss of water. Thus, a plant on a dry, sunny day might have some of its stomata closed to prevent desiccation.

Although the loss of excessive amounts of water through transpiration can be harmful to a plant, it is transpiration that allows for the movement of water and dissolved solutes throughout the plant, moving water upwards and against gravity. Water moves from an area of high **water potential** to an area of low water potential. Thus, when water evaporates from a leaf, it creates lower water potential there, and water will move through osmosis upward toward the leaves. The TACT mechanism (transpiration, adhesion, cohesion, tension)

> **DID YOU KNOW?**
> Trees don't get much taller than 100 meters because of a mathematical relationship governing leaf size and tree height. California redwoods (Sequoia sempervirens) are the tallest tree species on Earth, and the tallest recorded one is approximately 115 meters.

enables water and nutrients to travel from the soil, into the xylem, up the trunk, and into the leaves, even if the plant is a 100-meter tall tree like a redwood.

## Science Practices

There are two possible methods you could have used to measure the transpiration rate. The first was to assemble a potometer, which consists of a plant cutting, clear plastic tubing, and a calibrated pipette. The tubing and pipette were filled with water, and the plant was placed on the open side of the tubing. As transpiration occurred, it drew water from the tubing and the amount of water lost by the plant was measured using the calibrations on the pipette. Alternately, you may have attached your potometer to a pressure probe and measured transpiration through a change in pressure.

The second method, called whole plant transpiration, involved putting the roots of a small potted plant into a plastic food storage bag and leaving the leaves exposed. The plant was massed on day one and then again for several days after being exposed to the experimental conditions. This created a closed system so that any mass lost by the plant was due to transpiration—water loss from the leaves.

## Data Analysis

For the stomatal peel, it is likely you investigated the number of stomata on one plant and compared it to other plants based upon their habitat. Refer to your lab manual for a table listing some average stomata number on various plants. Typically, plants living in moist habitats, or receiving more sunlight, or living in areas with lower carbon dioxide concentrations have more stomata.

Next, you determined the rate of transpiration. You shared your data with other members of the class who tested different environmental variables and graphed the class data. The way you calculated rate of transpiration depends upon the method used in your experimental setup. For this review book, we will use data that could be generated using the whole plant method. This involved calculating the percent mass change of plants on a daily basis. Then the percent change was divided by four days to determine transpiration rate. Sample data (reported in percent mass change) and a graph are shown below.

| Condition | Monday | Tuesday | Wednesday | Thursday | Friday | Rate (% change/day) |
|---|---|---|---|---|---|---|
| Control | 0 | 10.0 | 14.9 | 23.1 | 31.7 | 7.9 |
| Light | 0 | 15.8 | 30.8 | 43.7 | 50.5 | 12.6 |
| Fan | 0 | 14.6 | 26.5 | 36.0 | 37.4 | 9.3 |
| Mist | 0 | 0.7 | 1.2 | 1.9 | 2.4 | 0.6 |

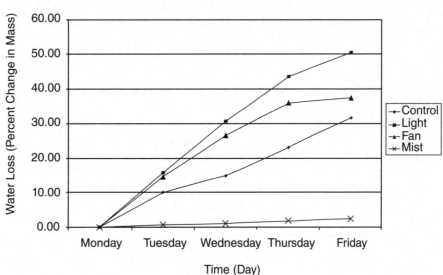

For simplicity sake, the data above does not take into account leaf surface area. You may have measured transpiration rate based upon surface area, so the next calculation you would have done is of leaf surface area. There are several ways to do this, but the most common is to trace each leaf onto a piece of graph paper (where 4 blocks equal 1 $cm^2$), count the number of squares covered, and divide by four to get a value measured in $cm^2$. If you used this method, then once you calculated your individual rate of transpiration, you would have divided that rate by the surface area of all the leaves in your whole plant.

## Conclusions

The results of this lab depend upon the exact experimental set-up used in your classroom. The main ideas you should have learned from this lab are:

- Stomata are visible under a microscope. Their density on a leaf surface is determined by the plant's habitat: higher moisture, higher temperature, and decreased carbon dioxide cause higher densities. Essentially, environmental pressures over time determined the evolution of stomata number on varying plant species.

- Transpiration rate increases with higher temperature, increased light, and wind. Thus, plants maintain homeostasis in varying environmental conditions by regulating the rate of transpiration. Opening and closing their stomata will regulate water loss through the leaves.

# Lab 12: Fruit Fly Behavior

## Concepts

In order to survive, animals must respond appropriately to their environment. The purpose of this lab was to examine animal behavior in a changing environment, specifically taxis. The College Board lab suggests using fruit flies (*Drosophila melanogaster*) because they are small, well-understood, easily manipulated animals used in many biology labs. However, your teacher might have chosen another organism, such as the pill bug, for this lab.

Orienting behaviors, including kinesis and taxis, are exhibited by animals as they move toward their favored environments. **Kinesis** is random movement that does not orient an animal in any particular direction, for example, when an animal is exposed to light and moves randomly in all directions. **Taxis**, however, occurs when an animal moves either towards or away from a stimulus, for example, when an animal moves away from light. Movement towards the stimulus is considered positive, while movement away from the stimulus is considered negative. The following table gives several examples of types of stimulus and the associated taxis.

| Stimulus | Taxis |
|---|---|
| Light | Phototaxis |
| Contact/touch | Thigmotaxis |
| Heat | Thermotaxis |
| Chemicals | Chemotaxis |
| Gravity | Geotaxis |
| Water | Hydrotaxis |
| Electricity | Galvanotaxis |
| Magnetic field lines | Magnetotaxis |

## Science Practices

This lab let you design your own animal behavior experiments, so methods varied. If you used fruit flies in this lab, then you probably learned to determine the sex of individual flies. Typically, males are smaller than females and their abdomens are much darker in color. You might have used this information to design an experiment comparing the preferences of environmental conditions between males and females.

To determine animal taxis, you probably constructed a choice chamber out of two plastic water bottles. By setting up differing conditions on either end of the choice chamber, you could quantitatively measure animal behavior by counting the number of flies (or other animal) on either end of the chamber. Before using the choice chamber in an experimental condition, you should have tested it with distilled water to be certain that no other variables were affecting your chamber. For example, if one side of the chamber was receiving more light, then it might have skewed your results because fruit flies prefer lighter conditions.

## Data Analysis

You should have collected numeric data about the location of fruit flies in the choice chamber. Because student experiments varied, hypothetical data for taxis behaviors of 30 fruit flies choosing between mayonnaise and mustard will be used and is presented in the table below. For each trial, the flies were allowed to move for two minutes and the number on each side of the choice chamber was recorded at the end of the two minutes.

| Trial | Number of flies on mustard side | Number of flies on mayonnaise side |
|---|---|---|
| 1 | 18 | 12 |
| 2 | 20 | 10 |
| 3 | 19 | 11 |
| 4 | 22 | 8 |
| 5 | 21 | 9 |
| 6 | 23 | 7 |
| 7 | 27 | 3 |

(*Continued*)

(*Continued*)

| Trial | Number of flies on mustard side | Number of flies on mayonnaise side |
|---|---|---|
| 8 | 21 | 9 |
| 9 | 24 | 6 |
| 10 | 26 | 4 |
| Average | 22.1 | 7.9 |

It may seem obvious from the data that the fruit flies chose mustard over mayonnaise, thus demonstrating positive chemotaxis for mustard. However, in scientific data analysis, we can't just look at the numbers and say with any certainty that an outcome is true. We must analyze the data statistically to see if there is a significant difference. For this lab, the chi-square statistical test was used to analyze data. The use of the chi-square requires the generation of a null hypothesis. For this lab, the null hypothesis was: If the fruit flies do not have a preference for mustard or mayonnaise, then there will be an even number of flies on each side of the choice chamber after two minutes. For the test of 30 fruit flies, you would expect 15 flies on each side of the choice chamber. The chi-square formula is $x^2 = \sum (o-e)^2/e$ and the calculations are shown in the table below.

|  | # Observed | # Expected | (o−e) | (o−e)² | (o−e)²/e |
|---|---|---|---|---|---|
| Mustard | 22.1 | 15 | 7.1 | 50.41 | 3.36 |
| Mayonnaise | 7.9 | 15 | 7.1 | 50.41 | 3.36 |
|  |  |  |  | Total | 6.72 |

Because there were two sides (mayonnaise and mustard), there was one degree of freedom for your calculation (df = number of groups − 1). The probability value was 0.05, so with 1 df, the critical value was 3.84. If your calculated chi-square was greater than 3.84, you rejected your null hypothesis. Because the calculated value of the sample data was 6.72 and greater than 3.84, the null hypothesis was rejected.

# TEST TIP

The chi-square formula will be provided for you on the formula sheet. Although you don't need to memorize it, you should be able to apply it to a given set of data. There will certainly be at least one chi-square calculation on the test. The formula sheet will also have a table of chi-square values.

## Conclusions

Animals respond to changes in their environment. There are some general trends of fruit fly behavior you may have observed in this lab.

- Positive phototaxis—Adult fruit flies are attracted to bright light (although larvae move away from light).

- Negative geotaxis—Fruit flies climb up their chambers away from gravity.

- Positive chemotaxis—Fruit flies are attracted to rotting or fermenting fruits. These fruits provide both a food source and a place to lay eggs so the larvae will have food. It is not the fruit itself that attracts the flies; instead, they are attracted to any foods with vinegar or alcohol. This is because rotting fruits produce alcohol or vinegar as products of decomposition. Thus, if you tried a substance like mustard, the fruit flies should have shown positive taxis to it because of the high vinegar content.

> **DID YOU KNOW?**
>
> Many AP Biology teachers end up with fruit fly infestations in their classrooms after conducting this lab. One trick to catching them is to fill a small beaker with apple cider vinegar, cover it with plastic wrap, and poke a hole in the wrap. The flies are attracted to the vinegar and will fly into it, however, most cannot get out and will be captured in this vinegar "trap."

# Lab 13: Enzyme Action

## Concepts

Chemical reactions in living organisms are sped up by **enzymes**, biological catalysts which lower the activation energy needed for a reaction to begin. An enzyme works by binding the substrate at an active site and forming an enzyme-substrate complex. This complex is energetically less favorable than the substrate alone, so the chemical reaction will proceed. A summary of the way an enzyme works is in Figure 23-3.

Figure 23-3. A General Enzyme Catalyzed Reaction

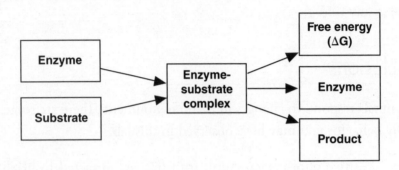

Most enzymes have optimal conditions for working; for example, the lactase used to digest lactose sugar in your small intestines works best at a pH of 7 and a temperature of 37 °C, while an enzyme found in a thermophilic (heat-loving) bacterium would require a much higher temperature such as 50 °C. Varying conditions away from the optimum can cause the protein to denature, or unfold, from the three-dimensional structure that allows it to function. Any change in the shape of the active site where the substrate binds contributes to a loss of enzyme function. In this lab, you tested the function of the enzyme peroxidase under varied conditions, such as pH or temperature. Peroxidase is found in many living organisms, is produced in the peroxisome, and breaks down hydrogen peroxide into water and oxygen gas. This equation can be summarized as:

$$2H_2O_2 + \text{peroxidase} \longrightarrow 2H_2O + O_2 \text{ (gas)} + \text{peroxidase}$$

## Science Practices

The purpose of this lab was to determine how the rate of peroxidase action varied with different environmental conditions. There are multiple ways to measure enzyme activity; it depends upon your teacher which method you used. Essentially, all methods work by measuring the production of oxygen, with a greater production of oxygen indicating a greater rate of reaction. The three most common methods are:

1. Qualitative analysis using a colored indicator called guaiacol which gets darker when more oxygen is present. This method used a color pattern to estimate how much oxygen was produced. Essentially, darker colors indicated more oxygen production.

2. Quantitative analysis using the guaiacol indicator and a spectrophotometer or colorimeter to measure absorbance. Darker colors had a higher absorbance and indicated more oxygen production.

3. Quantitative analysis using a pressure probe or oxygen sensor. An oxygen sensor directly measured oxygen production. Alternately, increased oxygen production would cause an increase in pressure inside the test tube, which could be measured with a pressure probe.

## Data Analysis

Although there were several ways to measure the rate of the enzymatic reaction, this review will present data quantitatively as absorbance (please note absorbance is unitless). The most common conditions tested in this lab are temperature, pH, enzyme concentration, and substrate concentration, thus results are presented for these four conditions only.

1. Effect of Temperature on Enzyme Activity—Absorbance data was collected after the reaction had proceeded for three minutes. Sample data and a graph are shown below. Peroxidase worked best at a temperature range of 35–45 °C.

| Temperature | 5 °C | 15 °C | 25 °C | 44 °C | 55 °C | 70 °C | 100 °C |
|---|---|---|---|---|---|---|---|
| Absorbance | 0.11 | 0.18 | 0.26 | 0.32 | 0.27 | 0.17 | 0 |

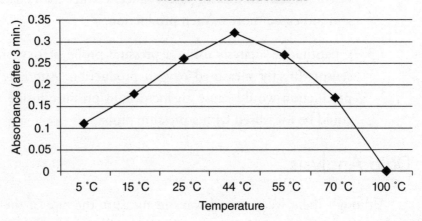

2. Effect of pH on Enzyme Activity—Sample data for absorbance and a graph are shown below. Peroxidase worked best at a pH range of 5 to 6.

| pH | 3 | 5 | 6 | 7 | 8 | 10 |
|---|---|---|---|---|---|---|
| Absorbance | 0 | 0.7 | 0.4 | 0.2 | 0.1 | 0 |

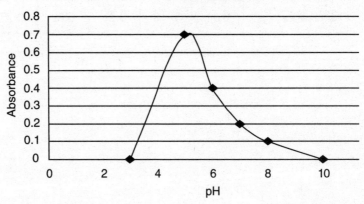

3. Effect of Varying Enzyme and Substrate Concentrations on Enzyme Activity— The baseline was generated using 0.3 ml of 0.1 percent hydrogen peroxide and 1.5 mL of peroxidase. Absorbance data was collected every minute for five

minutes after the reaction started. Sample data and a graph are shown below. The rate of the reaction could be calculated from the slope of the line. For example, the baseline rate was: 0.32 − 0.07/4 minutes = 0.063/minute and the rate of 2x enzyme concentration was: 0.48 − 0.14/4 minutes = 0.085/minute. Thus, the rate of the reaction increased with an increase in enzyme concentration.

|  | 1 minute | 2 min. | 3 min. | 4 min. | 5 min. |
| --- | --- | --- | --- | --- | --- |
| Baseline | 0.07 | 0.13 | 0.19 | 0.29 | 0.32 |
| 0.5 enzyme | 0.02 | 0.08 | 0.10 | 0.13 | 0.16 |
| 2x enzyme | 0.14 | 0.25 | 0.38 | 0.44 | 0.48 |
| 0.5 substrate | 0.07 | 0.14 | 0.19 | 0.28 | 0.30 |
| 2x substrate | 0.08 | 0.15 | 0.21 | 0.30 | 0.33 |

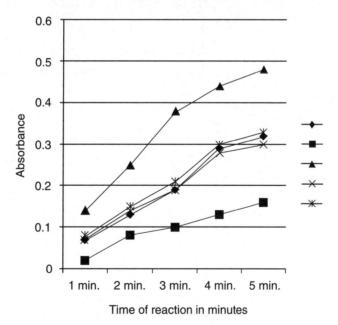

**The Effect of Substrate and Enzyme Concentration on Peroxidase Activity as Measured with Absorbance**

## Conclusions

As with most enzymes, the activity of peroxidase is optimal at certain environmental conditions. These conditions are based upon the living conditions of the organism using the enzymes. For example, the peroxidase used in this experiment was extracted from turnips. It worked better at a pH of 5 to 6 because soil pH hovers in that range. Optimum temperature is usually determined by the temperature at which the enzyme denatures. Slower reaction rates at lower temperatures are due to lower kinetic energy of the molecules so enzyme and substrate don't collide as often in order to react. Finally, enzyme concentration was a limiting factor on reaction rate. When the enzyme was doubled, that allowed the reaction rate to increase because more enzyme molecules were available to catalyze the breakdown of hydrogen peroxide.

**Time for a quiz**
- Review strategies in Chapter 2
- Take Quiz 7 at the REA Study Center
  *(www.rea.com/studycenter)*

**Take Mini-Test 2**
on Chapters 14–23
Go to the REA Study Center
*(www.rea.com/studycenter)*

# Practice Exam

Also available at the REA Study Center *(www.rea.com/studycenter)*

> This practice exam is available at the REA Study Center. Although AP exams are administered in paper-and-pencil format, we recommend that you take the online version of the practice exam for the benefits of:
>
> - Instant scoring
> - Enforced time conditions
> - Detailed score report of your strengths and weaknesses

# Practice Exam
## Section I
### Part A

(Answer sheets appear in the back of the book.)

**TIME:** 90 minutes
63 multiple-choice questions

**Directions:** Each of the questions or incomplete statements below is followed by four suggested answers or completions. Select the answer that is best in each case. When you have completed Part A, continue on to Part B.

1. In order to determine the structure of DNA, Watson and Crick drew on the experimental work of a number of scientists. Erwin Chargaff was a biochemist who studied the nitrogenous bases of nucleic acids. Below is data produced by Chargaff that reports on the nucleotide base composition of DNA from various organisms. What is an important inference drawn from this data?

| Percentage of Nucleotide Bases in Various Organisms | | | | |
|---|---|---|---|---|
| Species | A | G | C | T |
| *Zea mays* (corn) | 26.8 | 22.8 | 23.2 | 27.2 |
| *Octopus vulgaris* (octopus) | 33.2 | 17.6 | 17.6 | 31.6 |
| *Gallus gallus domesticus* (chicken) | 28.0 | 22.0 | 21.6 | 28.4 |
| *Homo sapiens* (human) | 29.3 | 20.7 | 20.0 | 30.0 |
| *Triticum aestivum* (wheat) | 27.3 | 22.7 | 22.8 | 27.1 |
| *Saccharomyces cerevisiae* (yeast) | 31.3 | 18.7 | 17.1 | 32.9 |
| *E. coli* (bacteria) | 24.7 | 26.0 | 25.7 | 23.6 |

(A) All organisms have roughly equal amounts of each of the four nucleotide bases.

(B) The percentage of nucleotide base A is always higher than the percentage of nucleotide base G.

(C) There is a regularity in the ratios of bases A:T and C:G within each organism.

(D) Since the nucleotide base differences between such varied organisms as octopus and yeast are so small, nucleotide bases must not be an important contributor to the structure of DNA.

2. HeLa cells are immortal cancer cells taken from a cervical cancer patient named Henrietta Lacks. These cells show extra DNA due to the presence of viral DNA, as well as other chromosomal abnormalities. In normal cells, when DNA damage occurs that cannot be repaired, the p53 protein activates "suicide" genes that produce proteins that do which of the following?

(A) The p53 protein initiates proteins that decondense chromosomes, thereby stopping the cell cycle.

(B) The p53 protein initiates apoptosis by activating genes that produce nucleases and proteases that will digest the cell—effectively killing it.

(C) The p53 protein activates proteins that destroy the mitotic spindle, thus halting the cell cycle and, ultimately, the life of the cell.

(D) The p53 protein activates proteins responsible for shortening the telomeres, thus hastening the end of the cell's ability to continue through mitosis.

3. The gene density of an organism is a measure of how many genes are present per million base pairs. There is great variation in the gene density of different organisms, and there does not seem to be a correlation between how complex an organism is and its gene density. For example, humans have about 6,000 million base pairs and fewer than 21,000 genes, whereas the plant Arabidopsis only has about 125 million base pairs but has about 25,000 genes. Which of the following processes is probably an important mechanism for enabling a complex organism to get by on so few genes?

(A) Alternative splicing of a given transcript means that one transcript can produce different proteins.

(B) A protein is very malleable; thus, it can carry out many different functions within an organism.

(C) Complex organisms have more sophisticated regulatory mechanisms so fewer different proteins are necessary to carry out a given process than in other organisms.

(D) The number of genes per million base pairs, the gene density, is not fully understood. It is probable that scientists will find that there is more protein-coding DNA within complex organisms than currently measured.

4. Which of the following descriptors best compares mitosis to meiosis, respectively?

   (A) Mitosis is preceded by one round of replication; meiosis is preceded by two rounds of replication.

   (B) Mitosis requires two cell division sequences; meiosis requires one cell division sequence.

   (C) Mitosis halves the number of chromosomes present in the cell; meiosis retains the number of chromosomes present in the cell.

   (D) Mitosis produces two cells identical to the parent cell; meiosis produces four cells, each different than the parent cell.

5. Bacteria have their genes for a particular metabolic pathway organized sequentially on their circular chromosome, and they are able to rapidly adjust their enzyme production to suit the availability of nutrients. Because the genes for a metabolic pathway are sequential, once a pathway is activated, a single mRNA transcript of the whole set of genes is produced. Below is a schematic of a *lac* operon, an inducible operon found in *E. coli*. This *lac* operon is active when the nutrient lactose is present in the bacterial environment. Given this schematic and information, what would be the result of a mutation in the regulatory gene?

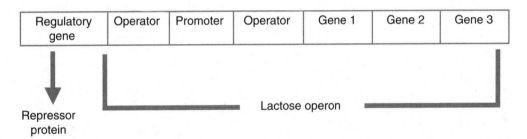

   (A) The lactose genes would be continuously transcribed.
   (B) The lactose genes would not be transcribed.
   (C) The lactose genes would be transcribed only if lactose is present.
   (D) The lactose genes would be transcribed only if lactose is absent.

**Question 6.**

The Punnett square represents possible patterns of inheritance in dihybrid crosses. Brown eyes (B) are dominant and blue (b) are recessive. Straight hair (S) is dominant and curly hair (s) is recessive.

|    | BS | Bs | bS | bs |
|----|----|----|----|----|
| BS | A  | B  | C  | D  |
| Bs | E  | F  | G  | H  |
| bS | I  | J  | K  | L  |
| bs | M  | N  | O  | P  |

If organisms of type J and type O are crossed, what fraction of the offspring would be heterozygous for both traits?

(A) $\dfrac{1}{16}$

(B) $\dfrac{3}{16}$

(C) $\dfrac{1}{4}$

(D) $\dfrac{7}{16}$

**Questions 7–8.**

In 1981, Francisco Nottebohm hypothesized that adult male canaries exhibit seasonal neuroplasticity by growing new neurons during breeding season in areas of the brain responsible for producing singing activity. Examining the canary brains in the spring and fall showed that there were dramatic changes in the canary brain song centers, supporting Nottebohm's hypothesis. Further experiments showed that if female canaries were given testosterone, their song centers enlarged as well. Finally, Nottebohm and S. A. Goldman injected females with testosterone that also included radioactively labeled thymidine nucleotides. The labeled thymidine would be incorporated into DNA as replication occurred before new cells formed. The results of this experiment showed many neurons in the birds' song centers containing the labeled thymidine, indicating they were newly-formed neurons.

7. What is the signaling molecule that generates the seasonal neuron formation in this series of experiments?

    (A) Thymidine

    (B) Testosterone

    (C) Estrogen

    (D) The experimental description gives no evidence for a signaling molecule.

8. What is the correct description of the action of this signaling molecule?

    (A) The signaling molecule binds to a G-protein-coupled receptor which activates GTP and a second messenger is activated.

    (B) The signaling molecule binds to a receptor tyrosine kinase, dimerization occurs, and each phosphorylated tyrosine activates a separate cellular response.

    (C) The signaling molecule passes through the cell membrane, binds to an intracellular receptor, and functions as a transcription factor to regulate gene expression.

    (D) The signaling molecule passes through the cell membrane, binds to an intracellular receptor, and initiates a cascade of "suicide" proteins within the cell.

9. What is the origin of a secretory vesicle?

    (A) The nuclear envelope buds off to form the secretory vesicle.

    (B) The lysosome buds off to form the secretory vesicle.

    (C) The Golgi apparatus buds off to form a secretory vesicle.

    (D) The cell membrane buds off to form a secretory vesicle.

**Question 10.**

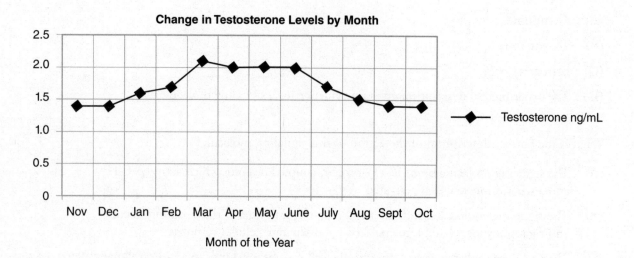

According to the graph, what is the proximate cause for the increase in testosterone levels in canaries?

(A) Temperature

(B) Day length

(C) Proximity of females

(D) The presence of thymidine

**Questions 11–12.**

The immune's system ability to recognize a wide range of antigens depends on the generation of a wide variety of different receptors found on cells of the immune system. The huge variety of receptors that is found in vertebrates is accomplished by gene rearrangement. For example, adenosine deaminase, one of a set of enzymes referred to as recombinases, randomly links three DNA segments, a V (for variable) segment, a J (for joining segment), and a C (for constant) segment to make a functional light or heavy chain on an immune cell receptor.

11. How would an individual with a mutation in the adenosine deaminase gene be affected?

   (A) The affected individual would make defective macrophages.

   (B) The affected individual would make defective B cells.

   (C) The affected individual would make defective neutrophils.

   (D) The affected individual would make defective red blood cells.

12. A light chain of an immune cell receptor is coded for by an immunoglobulin (Ig) gene. If an Ig gene has 70 different $V_L$ regions, 5 different $J_L$ regions, and 2 different $C_L$ regions, how many different light-chain receptor sites could be formed from this gene?

    (A) 10
    (B) 350
    (C) 700
    (D) $1.1 \times 10^5$

**Questions 13–15.**

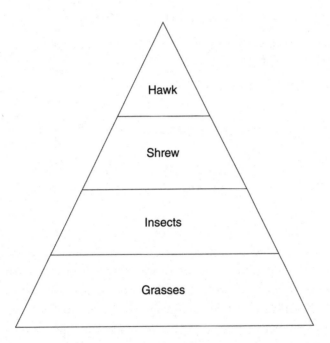

13. The figure above represents a simplified pyramid of trophic levels that exist for the organisms listed. On average, in a stable environment, 10% of the total energy from one level is available to the next trophic level. If the grasses have a productivity of $3.0 \times 10^3$ kcal/m²/year, what is the approximate area of grassland necessary to support one actively feeding red tail hawk, which requires about 174 kcal/day to thrive?

    (A) 17.2 m²
    (B) 172 m²
    (C) $2.1 \times 10^4$ m²
    (D) $6.4 \times 10^4$ m²

14. If DDT, an insecticide that persists in an environment and in organisms, were used to control the insects in this system, which of the following would be a logical consequence?

    (A) The grasses would show the highest accumulation of DDT per unit mass.

    (B) The insects would show the highest accumulation of DDT per unit mass.

    (C) The shrews would show the highest accumulation of DDT per unit mass.

    (D) The hawk would show the highest accumulation of DDT per unit mass.

15. Over time, the DDT ceased to be effective at controlling the insect population. Which of the following is the most likely reason for the drop in DDT effectiveness?

    (A) The insects developed a resistance to the DDT.

    (B) The initial population of insects was genetically diverse and at least some of the insects were resistant to DDT.

    (C) The DDT induced mutations and, in at least some of the insects, the mutation was resistant to DDT.

    (D) A population of resistant insects immigrated into the area bringing their DDT-resistant genes.

**Questions 16–17.**

In the 1930s, the U.S. Department of Agriculture conducted a series of experiments to determine the action spectrum for light-induced germination of lettuce seeds. Groups of prepared lettuce seeds (the seeds had been soaked in water) were exposed to one wavelength of light for a very short period of time, then placed in the dark for two days. The scientists then counted the number of germinated seeds in each group. They found that red light of wavelength 660 nm increased the germination percentages maximally, whereas exposure to far-red light (740 nm) most inhibited germination. When experimenters subjected prepared lettuce seeds to red light followed by far-red light, germination rates again dropped. But when they subjected the seeds to far-red followed by red, germination rates rose.

16. What is one conclusion that can be made from this series of experiments?

    (A) The effects of red and far-red light are reversible in the germination of lettuce seeds.

    (B) Lettuce seeds will not germinate if exposed to far-red light.

    (C) Red light is the single wavelength that will initiate germination in lettuce seeds.

    (D) Germination of lettuce seeds is independent of exposure to light. Germination depends on soaking the lettuce seeds and placing them in the dark for an appropriate period of time.

17. Further studies done much later than the germination experiment showed that seed germination occurred because of the presence of photoreceptors on proteins embedded in the surface of plant cell membranes. These photoreceptors were found to exist in two isomeric forms. The Pr form responded to red light exposure and converted to the Pfr form. The Pfr form responded to far-red light and reverted to the Pr form. Based on this information, which form of the photoreceptor is the active form, that is, the form that tranduces the light signal to the seed cells to initiate germination?

    (A) Both the Pr and Pfr forms transduce the signal to initiate germination.
    (B) The Pr form of the photoreceptor transduces the signal to initiate germination.
    (C) The Pfr form of the photoreceptor transduces the signal to initiate germination.
    (D) Both the Pr and Pfr forms are active, but the red light warmed the seeds more than the far red.

**Questions 18–19.**

Robert Paine, a biologist, observed that the number of species in an ecosystem decreased when the number of predators present decreased. To investigate this phenomenon, Paine conducted a removal experiment along North America's west coast. Paine put sea stars (*Pisaster ochraceus*) along with its usual prey, a mix of mussels, barnacles, limpets, and chiton, in control plots typical of the intertidal zone. In the experimental intertidal plots, Paine removed the sea stars. Over the year of the experiment, Paine saw a series of changes in the species composition of the experimental plots. After a year, the number of species in the experimental plots was reduced from 15 to 8 invertebrate species, with mussels being the most numerous species.

18. Which one of the following is the most likely hypothesis for this experiment?

    (A) Removal of predators allows beneficial species populations to flourish within a community.
    (B) The least numerous consumer is the most important member of a community.
    (C) Some consumers are critical in maintaining species diversity within communities.
    (D) One species of invertebrates will out-compete other populations of invertebrates.

19. In this intertidal community, the sea stars can be described as a(n)

    (A) adaptive species
    (B) keystone species
    (C) stabilizing species
    (D) indicator species

20. In most mammals, there are many anatomical structures that have adaptations that serve to increase the surface area for chemical reactions and exchanges between the structure and its environment. Which of the following is an example of surface area adaptations?

    (A) The large size of the plant cell vacuole
    (B) The round shape of the cell nucleus
    (C) The cristae of the mitochondria
    (D) The ligand-gate receptors in the cell membrane

21. The following is a sequence of DNA, what will be the mRNA transcript that will result from this DNA template?

    $$3' - \text{TAC CCG AAA ACT} - 5'$$

    (A) 5' – AUG GGC UUU UGA – 3'
    (B) 3' – AUG GGC UUU UGA – 5'
    (C) 5' – ATG GGC TTT TGA – 3'
    (D) 3' – ATG GGC TTT TGA – 5'

**Questions 22–23.**

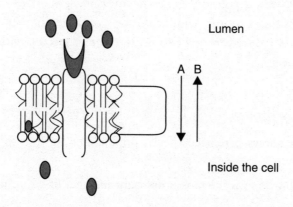

22. Fructose is one product of digestion that has to be moved into the cells lining the intestine. In the cross-section of an intestinal cell membrane above, the A arrow is pointing from the lumen (the passageway) of the small intestine into the cell lining it. The B arrow is pointing from inside the cell to the lumen. Movement of fructose into the cell will occur by which cell mechanism?

    (A) Facilitated diffusion

    (B) Osmosis

    (C) Active transport

    (D) Pinocytosis

23. Based on the molecules that make up the cell membrane, what can be inferred about the structure of fructose?

    (A) Fructose is incompletely broken down, thus too large to move across the cell membrane.

    (B) Fructose is polar and cannot move across the nonpolar interior of the lipid bilayer.

    (C) Fructose will be phagocytized in order to be taken into the cells.

    (D) Fructose is a liquid and must move into a cell by pinocytosis.

**Questions 24–26.**

In recent studies, digestive enzymes have been implicated in causing shock, a life-threatening condition that can follow a physical trauma such as a severe allergic reaction, hemorrhage, and massive infections. In cases of shock, blood pressure drops and disrupts blood flow to parts of the body including flow to the small intestine so that the protective mucosal lining is not maintained. Scientist Geert Schmid-Schönbein induced shock in rats with various treatments, then assessed intestinal tissue and the bloodstream for the presence of digestive enzymes. In subsequent experiments, Schmid-Schönbein induced shock in rats then injected enzyme blockers into one group and gave no treatment to another group of shock-induced rats. (*Science News*, Vol. 183, No. 4, Feb. 23, 2013)

24. What is a possible hypothesis for Schmid-Schönbein's first experiment?

    (A) The drop in blood pressure and subsequent reduced blood flow severely impairs a rat's immune response.

    (B) The drop in blood pressure and subsequent reduced blood flow causes the breakdown of the protective mucosal lining of the small intestine.

    (C) The drop in blood pressure and subsequent reduced blood flow causes waste products in the small intestine to build up, heightening the effects of the initial trauma.

    (D) The presence of digestive enzymes outside of the small intestine means that digestion has stopped and energy reserves in the body will drop to dangerous levels.

25. Schmid-Schönbein's data from the first experiment indicated a high count of digestive enzymes within the tissue of the small intestine and in the bloodstream of the body. What characteristic of digestive enzymes made their presence outside the small intestine dangerous?

    (A) The digestive enzymes could begin self-digestion of tissue and organ components of the organism.

    (B) The digestive enzymes act as antigens and establish an immune response.

    (C) The digestive enzymes are irreplaceable, thus the future health of the organism is impaired.

    (D) The conditions outside the small intestine will denature the enzymes and they will not be able to function properly.

26. If the active site of one of the most abundant digestive enzymes that escaped the confines of the small intestine was ionic in nature, which of the following combination of factors would most contribute to the successful choice of an enzyme blocker for the active site of this particular enzyme?

    (A) Use a relatively high concentration of blocking molecules that are nonpolar

    (B) Use a relatively high concentration of blocking molecules that are polar

    (C) Use a relatively low concentration of blocking molecules that are nonpolar

    (D) Use a relatively low concentration of blocking molecules that are polar

**Questions 27–29.**

After soaking overnight in water, plants were divided into four groups and the root balls wrapped in plastic. One group of plants was further misted with water and covered with plastic. One group was placed under grow lights that were on 24 hours a day. The next two groups were placed near an east-facing window. One of these groups was also placed 1 meter from a fan set to low to represent a gentle breeze.

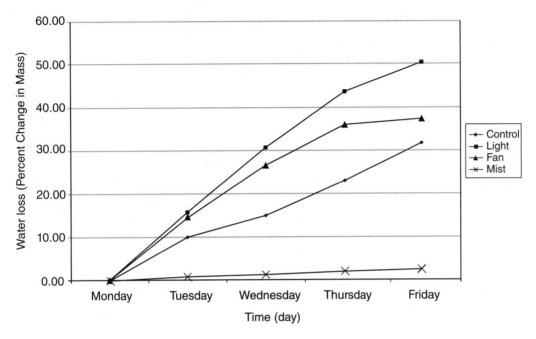

27. What is the average rate of water loss per day in plants kept in continuous light?

    (A) 7%/day
    (B) 10%/day
    (C) 13%/day
    (D) 20%/day

28. Guard cells surround the stomata of leaves and help control the opening and closing of the stomata. Potassium ions pumped into the guard cells make the cells hypertonic and water will move into the guard cells causing them to swell. If the guard cells are not turgid (swollen), they will flatten and this closes the stomata they surround. If the guard cells for all of these plants were deficient in $K^+$ ions, the data for which group of plants would be most significantly affected?

    (A) The control group would be most significantly affected.
    (B) The group kept in light would be most significantly affected.
    (C) The group subjected to the fan would be most significantly affected.
    (D) The group subjected to high humidity (mist) would be most significantly affected.

29. Where is the area of lowest water potential within the plants in this experiment?

    (A) Outside the stomata of the leaves in the plants exposed to the light

    (B) Outside the stomata of the leaves in the control plants

    (C) Outside the stomata of the leaves in the plants exposed to the fan

    (D) Outside the stomata of the leaves in the plants with the plastic covering (mist)

30. The medical community exploits the differences between prokaryotes and eukaryotes to develop antimicrobial drugs. For example, in eukaryotes, folic acid is ingested as a vitamin with food material and diffuses or is transported into a cell. In bacteria, folic acid cannot cross the cell wall and bacteria must synthesize it from para-aminobenzoic acid (PABA). The shape of the PABA is very similar to sulfa drugs, which inhibit the catalysis of glutamic acid to folic acid, a final step in folic acid production. What is the probable mode of action sulfa drugs have on this enzymatic pathway?

    (A) The sulfa drug denatures the enzyme that catalyzes the reaction of glutamic acid to folic acid.

    (B) The sulfa drug acts as a competitive inhibitor of PABA.

    (C) The sulfa drug acts as an allosteric inhibitor of the enzyme that converts glutamic acid to folic acid.

    (D) The sulfa drug acts as an activator of an alternative enzymatic pathway, shunting PABA out of folic acid production.

31. In 1648, Jan van Helmont put a carefully measured amount of dried soil in a large pot. He moistened the soil with rain water and planted a 2.3-kg willow shoot in the pot of soil. Van Helmont then covered the pot with a perforated iron plate to allow water and air in but keep as much dirt or debris out and planted the willow. After five years, van Helmont removed the willow tree, cleaned the dirt off of the willow roots in the pot, and massed the tree. Its new mass was 76.8 kg. Van Helmont dried the soil in the pot and re-massed it to find that it had lost only 57 g over the 5 years. Van Helmont concluded that the plant's increase in mass came from the water alone and not the soil. We now know that the increase in the mass of the plants is more the result of which of the following?

    (A) The hydrogen incorporated into the three-carbon sugars that result from photosynthesis

    (B) The minerals brought into the plant via transpiration

    (C) The $CO_2$ that is reduced in the process of photosynthesis

    (D) The combination of hydrogen incorporated into the three-carbon sugars and the minerals transported by transpiration and incorporated into plant molecules

## Questions 32–33.

The *lac* operon in bacteria is an inducible operon, that is, the genes for the metabolic pathway that can hydrolyze lactose are transcribed when lactose is present. When *both* glucose and lactose are present in a bacterial cell, transcription of the *lac* operon will be very slow. The cell preferentially hydrolyzes glucose because those enzymes are continuously present in the cell. If glucose levels fall in a cell, a small molecule called cyclic adenosine monophosphate (cAMP) accumulates and interacts with a regulatory protein, the catabolite activator protein (CAP), that attaches to DNA and acts as a transcription factor for the *lac* operon—increasing the rate of the operon transcription by increasing the attraction of RNA polymerase for the promoter region of the operon.

32. The lactose molecule activates transcription in which of the following ways?

    (A) Lactose acts as a feedback mechanism, shutting down the operon pathway.

    (B) Lactose interacts with the regulatory protein changing its conformation from active to inactive.

    (C) Lactose acts as a transcription factor enabling RNA polymerase to interact directly with the operon.

    (D) Lactose combines with cytoplasmic ribosomes and guides them to the growing operon transcript hastening translation.

33. The rate increase that occurs as a result of the drop in glucose qualifies this CAP/DNA interaction as which of the following?

    (A) The CAP/DNA interaction is an example of a negative feedback control mechanism.

    (B) The CAP/DNA interaction is an example of a positive feedback control mechanism.

    (C) THE CAP/DNA interaction is an example of a feedback system that is independent of its stimulus.

    (D) The CAP/DNA interaction is an example of an on/off control for the *lac* operon.

## Questions 34–35.

Cyanide acts as a poison by binding to an enzyme that is a component of electron transfer chains.

34. How will cyanide affect the production of ATP?

    (A) ATP production will decrease.

    (B) ATP production will increase.

    (C) ATP production will remain the same.

    (D) ATP will be dephosphorylated immediately after being formed.

35. Cyanide most directly disrupts the action of which of the following?

    (A) Cyanide displaces a coenzyme such as $NAD^+$.

    (B) Cyanide disrupts the phosphorylation of ADP to ATP.

    (C) Cyanide inhibits the oxidation of glucose.

    (D) Cyanide disrupts the formation of waste $CO_2$.

36. The following data table shows the water gain and loss by the kangaroo rat, a desert-dwelling mammal. What is the source of the metabolic water that balances the water loss in the kangaroo rat?

    | Water Gain (mL) | |
    |---|---|
    | Ingesting solids | 6.0 |
    | Ingesting liquids | 0.0 |
    | Metabolism | 60.0 |
    | **Water Loss (mL)** | |
    | In urine | 13.5 |
    | In feces | 2.6 |
    | By evaporation | 43.6 |

    (Source: *Biology: The Unity and Diversity of Life*, C. Starr and R. Taggart, 2004)

    (A) Condensation reactions make up the metabolic water gain.

    (B) Reduction of oxygen at the end of the electron transfer chain makes up the metabolic water gain.

    (C) Reduction of $NAD^+$ makes up the metabolic water gain.

    (D) Hydrolysis of macromolecules makes up the metabolic water gain.

37. Glycolysis is a highly-conserved process among living organisms. Which of the following best supports this statement?

   (A) Glycolysis is a process that is carried out in the cytoplasm of cells.
   (B) Glycolysis is a process that is carried out by virtually every living organism.
   (C) Glycolysis is a process that generates a large amount of usable energy for organisms.
   (D) Glycolysis is an anaerobic process.

**Questions 38–40.**

An experiment was conducted to determine the correlation between flowering and light exposure for a series of economically important plants. One hundred of each type of plant indicated in the table were given each of the light treatments indicated by 1 through 4. All other aspects of plant care were held constant. The data indicates the percent of plants that flowered in each group. All of the plants remained healthy throughout the treatment period.

| Light Treatment | Light/Dark Durations and Sequence |
|---|---|
| 1 | 15 hrs light/9hrs dark |
| 2 | 9 hrs light/15 hrs dark |
| 3 | 9 hrs light/7 hrs dark/1 hr light/7 hrs dark |
| 4 | 9 hrs light/4 hrs dark/1 hr light/10 hrs dark |

| Percentage that Flowered | | | | |
|---|---|---|---|---|
| Plant Type | 1 | 2 | 3 | 4 |
| Potatoes | 96 | 8 | 92 | 92 |
| Poinsettias | 7 | 98 | 8 | 7 |
| Spinach | 98 | 10 | 90 | 92 |
| Chrysanthemum | 7 | 97 | 33 | 32 |
| Tomatoes | 99 | 97 | 96 | 99 |
| Roses | 96 | 98 | 100 | 97 |

38. From the data shown, how many different types of plant responses to photoperiodism are present in this group of plants?

   (A) All of the groups show the same response to the differing light treatments applied.

   (B) There are two different flowering responses to the changing light treatments applied.

   (C) There are three different flowering responses to the changing light treatments applied.

   (D) There are four different flowering responses to the changing light treatments applied.

39. Which of the following plants showed the most sensitivity to the length of exposure to continuous darkness?

   (A) Potatoes
   (B) Poinsettia
   (C) Tomatoes
   (D) Roses

40. Based on the flowering pattern changes found in potatoes, which of the following most heavily influenced flowering patterns?

   (A) Potatoes are most influenced by the length of continuous sunlight to successfully flower.

   (B) Potatoes need about 9 hours of sunlight to successfully flower.

   (C) Potatoes are most influenced by the length of the night to successfully flower.

   (D) The potatoes successfully flower regardless of the influence of light. They have an internal biological clock that enables them to flower under the most appropriate conditions.

**Questions 41–42.**

Since 1990, scientists have been trying to harness the power of the immune system to destroy cancer. A study reported in February 2013 found that a partially disabled, genetically modified, cowpox virus infused into patients with advanced liver cancer extended their lives from an average of three to six months to fourteen months. The virus enters tumor cells, multiplies, and then the infected tumor cell blows apart. The cellular debris is phagocytized and an acquired immune response is triggered. (*Science News*, Vol. 183, No. 6, 3/2013)

41. In trying to hijack the immune system's machinery, which of the following actions were the scientists exploiting?

    (A) The scientists were trying to bring the virus in contact with cytotoxic T cells.

    (B) The scientists were trying to establish an inflammatory response to the cancer cells as this would trigger an immune response.

    (C) The scientists were trying to force an MHC cell to present a cancer cell fragment to a helper T cell.

    (D) The scientists were trying to use the action of cytokines to stimulate production of more viruses to invade more tumor cells.

42. On what characteristic does the immune system primarily depend for its success?

    (A) The immune system depends primarily on the action of phagocytosis and MHC presentation.

    (B) The immune system depends primarily on the interface between helper T cells and B cells.

    (C) The immune system depends primarily on the vast diversity and specificity of protein conformations.

    (D) The immune system depends primarily on the action of cytotoxic T cells.

43. Cancer is as common in dogs as in people, and their cancers are very similar to the human versions of the same cancers. In fact, many of the current treatments for cancer in humans started as corresponding treatments in dogs. Certain breeds of dogs exhibit increased risks of certain cancers as indicated in the table below. (*Science News*, Vol. 183, No. 5, 3/9/2013)

    | Breed | Cancer | Risk Increase |
    |---|---|---|
    | Chow chow | Gastric carcinoma | 10–20× |
    | Boxer | Mast cell tumors | 16.7× |
    | Scottish terrier | Malignant melanoma | 12.1× |
    | Great Dane | Osteosarcoma | 60.9× |

    Why do particular breeds exhibit such an increased risk of genetically-based cancers?

    (A) Selective breeding decreases genetic diversity.

    (B) Selective breeding increases the rate of mutations.

    (C) These breeds are exposed to the kinds of environments that increase the likelihood of mutations.

    (D) Because of their close association with humans, they are acquiring some of the same kinds of conditions as humans.

44. The majority of the ATP derived from cellular metabolism occurs in which of the following processes?

    (A) Substrate-level phosphorylation of ADP in glycolysis

    (B) Substrate-level phosphorylation of ADP during the citric acid cycle

    (C) The combined phosphorylation events from glycolysis and the citric acid cycle

    (D) The oxidative phosphorylation that results from the action of the electron transport chain

45. In a certain breed of flowers, red petals are incompletely dominant to white petals; the heterozygote is pink. Axial flowers are completely recessive to terminal flowers. If a red petal, axial flowered plant is bred to a pink flowered plant that is true-breeding for terminal flowers, what is the expected phenotypic ratio in their offspring?

    (A) The offspring will be 50% red petal, terminal flowers and 50% pink petal, terminal flowers.

    (B) The offspring will be 50% red petal, axial flowers and 50% red petal, terminal flowers.

    (C) The offspring will be 50% pink petal, axial flowers and 50% pink petal, terminal flowers.

    (D) The offspring will be 25% red petal, terminal flowers to 50% pink petal, terminal flowers to 25% white petal, terminal flower.

46. Mitosis often occurs when cells reach a certain size. This ensures which of the following?

    (A) The maximum amount of ribosomes is sufficient for protein production.

    (B) The surface area to volume ratio is adequate for exchange of materials to occur between the cell and its environment.

    (C) The mitochondria will be able to keep up with the energy demands of the cell.

    (D) That signal transduction can cross the cell membrane.

47. Bacteria are unicellular organisms but they can synthesize, release, and detect induction molecules in a process called quorum sensing. This allows the unicellular organisms to act in ways similar to multicellular organisms. Which of the following qualities most affect quorum sensing?

    (A) The presence of a mixture of microorganisms

    (B) The presence of oxygen in the environment of the microorganisms

    (C) The density of the microorganism population

    (D) The possession of Hfr (high-frequency of reproduction) plasmids

48. As a general rule, among endothermic animals, large bodies cool off more slowly than small bodies and, conversely, large bodies warm up more slowly than small bodies. What would you predict about the geographical location of a large mammal versus a small mammal?

    (A) A large mammal is likelier to be found in a warmer environment and a small mammal in a cooler environment.
    (B) A large mammal is likelier to be found in a cooler environment and a small mammal in a warmer environment.
    (C) Large and small mammals will be found in both locations in equal abundance.
    (D) In general, endothermic animals populate areas with overall mild temperatures.

49. Which of the following statements is true?

    (A) Diffusion is an active process that requires the cell to expend ATP.
    (B) Diffusion only occurs in living systems.
    (C) A cell placed in a salt solution will probably burst because of osmotic pressure.
    (D) When the concentration of solutes on both sides of a cell membrane is equal, the cell is said to be isotonic.

50. Which of the following does NOT introduce genetic variation into bacteria?

    (A) Conjugation
    (B) Transformation
    (C) Transduction
    (D) Antibiotics

51. Which of the following is NOT a part of the light reactions of photosynthesis?

    (A) The oxidation of water
    (B) The reduction of carbon dioxide
    (C) Generation of ATP by the electron transport chain
    (D) The reduction of $NADP^+$

52. The citric acid cycle takes place in the mitochondria. Which of the following statements is also true about the citric acid cycle?

    (A) The citric acid cycle will not take place in the absence of oxygen.
    (B) Oxygen is reduced in the citric acid cycle.
    (C) Carbon dioxide is reduced during the citric acid cycle.
    (D) The electron transport chain generates an electrochemical gradient.

53. Alcoholic and lactic acid fermentation occur at the end of glycolysis if oxygen is absent from the environment or if the organism is an anaerobic organism. What is the function of the conversion of pyruvic acid, generated by glycolysis, to either lactic acid or ethanol?

    (A) Lactic acid and ethanol are low toxic storage forms of waste for the organism.

    (B) Lactic acid and ethanol of these substances are available for further oxidation for the organism.

    (C) Lactic acid and ethanol accept electrons to regenerate $NAD^+$ for the organism.

    (D) Lactic acid and ethanol are readily removed from an organism as waste.

54. Which one of the following is NOT evidence for eukaryotic evolution by endosymbiosis?

    (A) All eukaryotic cells have mitochondria.

    (B) Mitochondria and plastids self-replicate.

    (C) Mitochondria and plastids have a single circular DNA molecule.

    (D) Mitochondria and plastids have a single lipid bilayer.

55. How is the body plan for most multicellular animals initiated?

    (A) Asymmetric distribution of maternal mRNA and proteins released into the egg act as transcription factors.

    (B) Asymmetric distribution of chromosomes which occurs after fertilization

    (C) Activation of the genome by fertilization

    (D) Sequential chromosomal activation after fertilization

56. Besides the plasmid vector, which list below contains correct ingredients necessary for successful plasmid recombination?

    (A) Restriction endonucleases, DNA, mRNA, nutrient media

    (B) Restriction endonucleases, mRNA, ligase, host bacteria, selective media

    (C) DNA, restriction endonucleases, ligase, selective media

    (D) DNA, proteases, selective media, mRNA

**Question 57–58.**

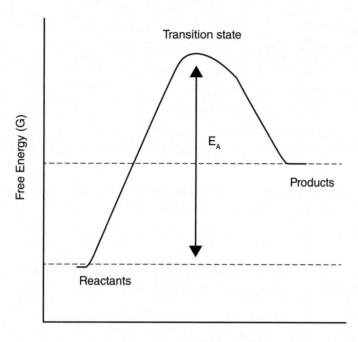

(Source: *http://en.wikibooks.org/wiki/Structural_Biochemistry/Endergonic_reaction*)

57. The energy changes in the graph above could represent the energy changes in which of the following?

    (A) The energy change indicated could result from the hydrolysis of a disaccharide.

    (B) The energy change indicated could result from the formation of a peptide bond.

    (C) The energy change indicated could result from the oxidation of a molecule of glucose.

    (D) The energy change indicated could result from the hydrolysis of ATP.

58. The area of energy change that is most affected by the presence of enzymes is the

    (A) initial energy of the reactants is raised

    (B) final energy of the products is lowered

    (C) overall change in energy of the reaction is lowered

    (D) energy of activation ($E_a$) is lowered

59. The net effect of the sodium-potassium pumps that reside in the membrane of a neuron and establish an action potential is best described by which of the following?

    (A) The sodium potassium pump uses ATP energy to maintain a zero charge balance between the internal and external environments of the neuron.

    (B) The sodium potassium pump uses ATP to pump three potassium ions out of the neuron for every three sodium ions pumped into the cell, setting up a membrane polarity.

    (C) The sodium potassium pump uses ATP to pump two potassium ions out of the neuron for every three sodium ions pumped into the cell, setting up a membrane polarity.

    (D) The sodium potassium pump uses ATP to pump two potassium ions into the neuron for every three sodium ions pumped out of the cell, setting up a membrane polarity.

60. Enzyme action within a chemical reaction, ligand-binding in a signal transduction pathway, and establishment of the acquired immune response all have which of the following in common?

    (A) All of these actions rely on the specificity of proteins.

    (B) All of these actions rely on ATP to supply the energy transformations.

    (C) All of these actions rely on the action of membrane proteins.

    (D) All of these actions rely on an electrochemical gradient established using ATP.

61. The initiation of an action potential requires which of the following?

    (A) A polarizing event

    (B) A depolarizing event

    (C) A minimum supply of ATP

    (D) A minimum supply of acetylcholine

62. What will characterize the succession that occurs after a farm has been abandoned?

    (A) Lichens will move into the abandoned area and colonize.

    (B) Neighboring trees will colonize the area.

    (C) Grasses and shrubs will colonize the area.

    (D) A cultivated area has experienced no essential change to its structure and there will be no succession.

63. The p53 gene codes for a protein that prevents cell growth and stimulates apoptosis in cells that have irreparable DNA damage. The p53 also increase the action of a protein, which blocks the formation of a critical cyclin dependent kinase. The ultimate effect on the cell is that the cell

    (A) will not complete mitosis

    (B) will disintegrate

    (C) will not replicate its DNA

    (D) will enter $G_0$ phase

# Section I
## Part B

**Directions:** The following six questions require numeric answers. Using the information provided with each question, calculate the correct answer and enter it on the grid-in answer sheet.

1. In studying genetics, a student conducted an experiment and crossed two pea plants that were heterozygous for plant height. The student generated a null hypothesis that half of the plants would be tall and half would be short. The student's data is presented in the table below.

   |  | Tall Phenotype | Short Phenotype |
   | --- | --- | --- |
   | Offspring | 572 | 428 |

   Using this data, calculate the chi-squared value for the null hypothesis. Give your answer to the nearest hundredth.

2. If blood circulating to a human's extremities is cooled to 18 °C and NOT re-warmed through countercurrent exchange, how many calories would a human lose every 24 hours (assuming the person remains in the same environment for the 24-hour period)? The core temperature of a human is 37 °C, and blood flow through the extremities is about 5 L/min. Blood is mostly composed of water, and the specific heat of water is 1 c/g °C (1000 c/l L °C).

3. The graph below was generated from an enzyme-catalyzed reaction. Use this graph to calculate the rate of the reaction between one and three minutes. Give your answer to the nearest tenth.

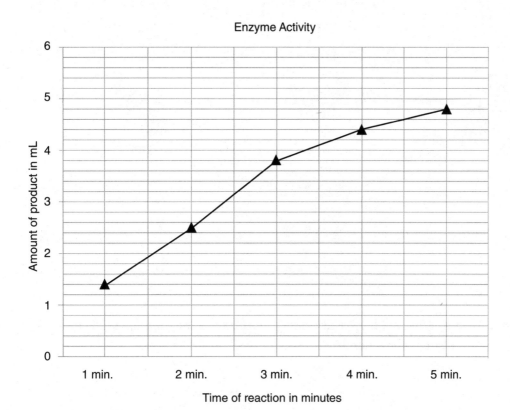

4. Rock pocket mice have two phenotypes for fur color. The light tan color is completely dominant to the dark brown color. In the year 2000, scientists counted the number of each color of mouse in a population and discovered that 84% of the population had light fur. Assuming the population is in Hardy-Weinberg equilibrium, what would the frequency of the dark fur allele be if scientists sampled the population today? Give your answer to the nearest tenth.

5. Refer to the food web of the flow of carbon in a forest ecosystem.

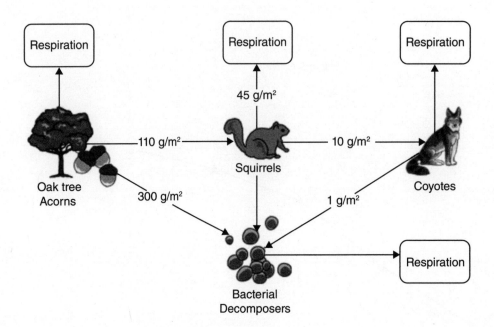

How much carbon in g/m² will be passed from the squirrels to the decomposers? Give your answer to the nearest whole number.

6. Parents having the following genotypes are crossed: AaBbCc × AabbCc. What is the probability of producing an offspring that shows the dominant phenotype for the A allele and the recessive phenotype for the B and C alleles? Give you answer as a fraction.

# Section II

**TIME:** 90 minutes, including a 10-minute mandatory reading period
**Questions:** 8

> **Directions:** Questions 1 and 2 are long free-response questions and should take about 20 minutes each to answer. The remaining six questions are short free-response questions and should take about 6 minutes each to answer. Be certain to carefully read and answer all parts of each question. Your answers must be written out; outlined or bulleted format is not acceptable.

1. Recognition of signals and regulation of cellular response is critical for organisms to respond to environmental changes and to maintain homeostasis. Examples of physiological systems in animals that carry out these processes include the immune, endocrine, reproductive, and nervous systems.

   Choose one specific plant signal and one specific animal signal and for each of the following:

   (a) **Identify** one signal that cells of the organism will receive and recognize and the physiological system affected (it does not have to be a system mentioned above).

   (b) For the signal identified in (a), **describe** how the signal is recognized by the cell.

   (c) **Explain** how the cell responds to the signal to maintain homeostasis.

   (d) **Discuss** the evolutionary advantage of each organism's response to the initial signal.

2. Two populations of plants are located around a large body of water. Population 1 lives in lowlands surrounding the lake. These lowlands have many other plant species and high humidity. Population 2 lives along the sides of a mountain overlooking the lake. These plants are low-growing shrubs that are similar in appearance, with a few small differences. However, their rates of transpiration are vastly different, as can be seen in the following data table. For many years, scientists believed these to be the same species of plants, but a small research group has proposed that they are actually two different species.

| Temperature °C | 20 | 23 | 25 | 27 | 30 |
|---|---|---|---|---|---|
| Population 1 transpiration rate (mmol/m²xsec) | 1 | 2 | 4 | 5 | 6 |
| Population 2 transpiration rate (mmol/m²xsec) | 2 | 4 | 6 | 7 | 9 |

(a) **Graph** the data and determine which plant species has the greatest rate of transpiration.

(b) **Propose** a hypothesis explaining why there might be a difference in transpiration rates between these two populations of plants. Provide evidence that would support your hypothesis.

(c) **Identify** two pieces of data that could be gathered from the existing populations that would support the hypothesis that these plants are the same species and that the transpiration rate differences are due only to environmental factors.

(d) **Explain** why each piece of data provides evidence.

3. Countercurrent exchange is a mechanism used by many organisms to regulate various physiological processes such as osmoregulation.

   (a) Choose an organism and a physiological process and **describe** how countercurrent exchange works in this organism.

   (b) **Explain** how countercurrent exchange contributes to the maintenance of homeostasis in the example you used in (a).

4. A population of rock pocket mice is being studied in the laboratory. These mice have two phenotypes for fur color: the lighter tan color is dominant to the dark brown color. Fur color is determined by a single, autosomal gene and the trait exhibits complete dominance. The scientist created the lab population by breeding 5 homozygous dominant tan mice with 5 brown mice. Data about the offspring was collected over seven generations; the data is presented below.

| Generation | Number of Individuals | | |
|---|---|---|---|
| | Tan | Brown | Total |
| 1 | 5 | 5 | 10 |
| 2 | 29 | 0 | 29 |
| 3 | 24 | 7 | 31 |
| 4 | 31 | 11 | 42 |
| 5 | 36 | 13 | 49 |
| 6 | 38 | 14 | 52 |
| 7 | 40 | 14 | 54 |

**Analyze** this data and **explain** if this population is in Hardy-Weinberg equilibrium or not.

5.  DNA replication is a highly-conserved process carried out by all living things; the ability to replicate defines living things and is why cells are termed living while viruses are nonliving.

    (a) **Explain** the process of DNA replication; include a comparison of the lagging versus leading strands in the answer.

    (b) **Describe** evidence from DNA replication that supports the idea that all living things are descended from a common ancestor.

6.  Feedback mechanisms are a common regulatory process used by the simplest prokaryotes up to the most complicated mammal to maintain homeostasis.

    (a) **Explain** how negative feedback mechanisms work, using an example in your explanation. **Discuss** how negative feedback is used to maintain homeostasis.

    (b) **Explain** how positive feedback mechanisms work, using an example in your explanation.

7.  Sickle cell anemia is a genetic disorder caused by a single substitution in the nucleotide sequence for hemoglobin. Hemoglobin is the protein that carries oxygen in red blood cells. The substitution switches the hydrophilic amino acid glutamic acid for the hydrophobic amino acid valine in one of the subunits of hemoglobin. This substitution changes the shape of normal red blood cells causing them to clog small blood vessels and severely impair their ability to carry oxygen. About one in ten African-Americans carries the allele for sickle cell anemia, a very high frequency for such a potentially harmful trait. Individuals receiving two alleles for sickle cell hemoglobin suffer from sickle cell anemia. Individuals who are heterozygous are phenotypically normal, although they express some sickle shaped red blood cells.

    (a) **Predict** why a harmful allele such as the sickle cell gene for abnormal hemoglobin would be maintained in the gene pool of a population.

    (b) Unlike in the sickle cell gene, some changes at the DNA level do not affect an organism's phenotype. **Explain** how this might happen.

8.  A large surface area to volume ratio is an important adaptation and recurs at all levels in the biological world. Using the mitochondria as an example, **explain** how increased surface area supports mitochondrial activity.

# Answer Key

## Section I

### Part A

1. (C)
2. (B)
3. (A)
4. (D)
5. (A)
6. (C)
7. (B)
8. (C)
9. (C)
10. (B)
11. (B)
12. (C)
13. (C)
14. (D)
15. (B)
16. (A)
17. (C)
18. (C)
19. (B)
20. (C)
21. (A)
22. (A)
23. (B)
24. (B)
25. (A)
26. (B)
27. (C)
28. (B)
29. (A)
30. (B)
31. (C)
32. (B)
33. (B)
34. (A)
35. (A)
36. (B)
37. (B)
38. (C)
39. (B)
40. (C)
41. (C)
42. (C)
43. (A)
44. (D)
45. (A)
46. (B)
47. (C)
48. (B)
49. (D)
50. (D)
51. (B)
52. (A)
53. (C)
54. (D)
55. (A)
56. (C)
57. (B)
58. (D)
59. (D)
60. (A)
61. (B)
62. (C)
63. (A)

### Part B

1. 20.74
2. 1.37
3. 1.2
4. 0.4
5. 55
6. 3/32

# Detailed Explanations of Answers

## Section I, Part A

1. **(C)**

    (A) is wrong because an examination of the data does not support the statement that there are roughly equal amounts of DNA bases. For example, in the octopus, there is 17.6% G but 33.2% A. (B) is not supported by the data. For example, note that in *E. coli*, there is 24.7% A, but there is more G at 26.0%. (C) is correct. Careful examination of ALL the data presented shows that in each organism, the percentage of A was very close to the percentage of T. For example, there were 26.8 to 27.2 in corn and 28.0 to 28.4 in the chicken. The same type of relationship was found between C and G. This regularity in his data was noted by Chargaff, and, coupled with other experimental data, helped Watson and Crick develop their model of DNA. (D) is also an incorrect statement. The data shows a high degree of molecular diversity that scientists had not been aware of until Chargaff did his experimentation.

2. **(B)**

    The p53 gene produces the p53 protein, which acts as a tumor suppressor. It can bind to CdKs to allow time for DNA repair, and it can activate proteins responsible for repairing DNA. If DNA is not reparable, as indicated in this question, p53 will activate apoptosis, cell suicide. (A), (C), and (D) may all contribute to cell death but they do not represent apoptosis, which is what the question is really asking about.

3. **(A)**

    Alternative splicing of a transcript has the potential to produce two or more proteins per gene. Proteins are not malleable (B) and, though the protein may carry out a different function in different cells, the protein itself does not change. There is no evidence that complex organisms (C) have a more sophisticated regulatory mechanism, and there is no evidence that a more sophisticated regulatory mechanism would require fewer proteins. (D) is simply not a correct statement.

4. **(D)**

Mitosis occurs in somatic cells and results in two daughter cells with the same number of chromosomes as the parent cell. Meiosis occurs in the gametogenic cells and results in four daughter cells. Because of crossover events in prophase I and independent assortment in meiosis I and II, each gamete is very likely different from the parent cell and has half the number of chromosomes.

5. **(A)**

The regulatory gene in an operon codes for the repressor protein that, when present, loops and binds the two operator genes and stops transcription of the *lac* operon. If the regulatory gene is mutated, a nonfunctional repressor protein will be produced so the *lac* operon will not be repressed; thus it will be continuously expressed.

6. **(C)**

The genotype of J is BbSs and this is crossed with O with a genotype of bbSs. The simplest way to solve this is to use the product rule. That is, the chance of Bb × bb giving heterozygous offspring is ½. The chance of Ss × ss giving heterozygous offspring is ½. The product rule states that the chance of two independent events (Bb and Ss) occurring simultaneously is the product of each individual chance: ½ × ½ = ¼.

7. **(B)**

(A) is incorrect because thymidine is a nucleotide found in all DNA and is not a signaling molecule. In this experiment, it was used as a label to indicate cell formation because as DNA is replicated prior to mitosis, the labeled thymidine would show up in areas of cell division. Estrogen (C) was never mentioned as one of the experimental variables. Testosterone (B) is specifically associated with seasonal neuron formation when the experimenters used testosterone to generate new neurons in female birds. (D) is incorrect because the description specifically states that testosterone is the probable signaling molecule.

8. **(C)**

(A) and (B) describe signal transduction; the signaling molecule would remain outside the cell. This kind of action is characteristic of polar signals. It is important to remember that steroid hormones generally *enter* the cell and function as transcription factors (C). Nothing in this experimental description talks about cell suicide (D).

9. **(C)**

   The Golgi apparatus is the organelle that receives proteins destined for secretion from the endoplasmic reticulum. When the protein is processed, a vesicle will bud off of the Golgi apparatus and will move to merge with the cell membrane and the protein will be released to the outside as the vesicle merges with the cell membrane.

10. **(B)**

    (B) can be inferred as the correct response because the graph of testosterone levels shows a direct relationship to increasing day length. Days begin to get longer from late December through June and that is the time of increased testosterone levels. (A) is incorrect because temperatures can vary widely between December and June but this graph does not vary widely. There is nothing in this graph that gives information about the presence of females (C) or the presence of thymidine (D), negating these as possible answers.

11. **(B)**

    (A), (B), and (C) are all cells of the immune system, but only B cells (B) and T cells (not a choice in the question) have receptor proteins on their surface that can bind a particular foreign molecule. Red blood cells (D) originate from the same stem cells as cells of the immune system, but do not function in fighting infection.

12. **(C)**

    Any one of the V regions can go with any one of the J or C regions. The possible combinations, then, is the product of the individual possibilities (a variation on the product rule). So the correct response is 70 × 5 × 2 (C).

13. **(C)**

    If 10% of the energy of a level is available to the next level, then 3 kcal/m²/yr is available to the hawk (1/1000 of the original 3000 kcal/m²/yr). To determine the area of grasses necessary to support a hawk, the energy required by the hawk is divided by the energy available per unit area:

    $$\frac{\text{Energy required}}{\text{Energy available / unit area}} = \frac{174\,\text{kcal}}{\text{day}} \times \frac{365\,\text{day}}{\text{y}} \times \frac{y\,(m^2)}{3\,\text{kcal}} = 2.1 \times 10^4\,m^2$$

14. **(D)**

    This is an example of biomagnification. The insects are sprayed and the DDT enters their bodies. This either slows them or kills them, and they are then eaten by the shrew at which point the DDT makes its way into the tissues of the shrew. Each shrew will eat many insects and, therefore, will have a higher per unit mass of DDT. Finally, the hawk will ingest several DDT-laden shrews and the relatively large amount of DDT will make its way into the tissues of the birds.

15. **(B)**

    Organisms do not develop a resistance (A) nor mutate directionally (C). There is no indication that a resistant population moved into the area (D). The likely scenario as described by Darwin is that the original population had variations, among them resistance to DDT. When the DDT-sensitive population was killed off, there was no competition and only the resistant insects were left.

16. **(A)**

    Maximum germination did occur after exposure to red light (C); however, the follow-up experiments showed that the action of red light could be reversed with far-red light making (A) a better conclusion. There was some germination with all of the seeds so (B) is not correct. (D), preparing the seeds by soaking them, was necessary for germination, but no results indicate that soaking alone would initiate germination.

17. **(C)**

    The experimental data presented indicates that only red light initiated germination. Since red light exposure converts Pr to Pfr, it can be concluded that the active form of this protein is the Pfr form (C). (A) and (B) are incorrect because the experiment indicated that exposure to far-red light *did not* initiate germination. This wavelength of light converts Pfr to Pr; thus, the Pr isomer is not active. (D) is not correct because there is no data regarding temperature presented in this experiment.

18. **(C)**

    In this ecosystem, the sea stars were not the most numerous organisms but, because they prey on mussels, they keep the mussel population in check so that this population does not out-compete all other invertebrate species present. There was no indication that certain species were more beneficial than others in this experiment (A), and it was not indicated in the experimental description whether the sea star was the least numerous organism (B). It is clear that the mussels out-competed other species, but this is only one experiment; further experimentation would be needed to determine if this is always the case (D).

19. **(B)**

    A keystone species is a single species whose niche is so significant that their removal can cause the collapse of an ecosystem. They are the "key" to the integrity of the system

(B). (A) and (C) are made up terms and (D), an indicator species, is one that acts as a "warning" of impending change in biodiversity.

20. (C)

    Branching, folding, or membranous extensions are ways of increasing surface area without appreciably changing the volume. The plant cell vacuole (A) and the nucleus (B) both have smooth surfaces without the characteristics of increased surface area. Ligand-gated receptors are very specific to their ligand and rely on specificity for their action NOT on increased surface area.

21. (A)

    The strand given is 3′ to 5′ and transcription occurs 5′ to 3′ so that rules out (B) and (D). (C) is incorrect because a transcript will have uracil bases, not thymine bases.

22. (A)

    Facilitated diffusion is characterized by movement of a solute down its concentration gradient as indicated by the higher number of solute molecules outside the cell versus inside the cell. Osmosis (B) is movement of water molecules with their gradient. Active transport (C) is movement of the solute from lower concentration to higher concentration. Pinocytosis (D) is the term used for the cell taking in large droplets of fluid material.

23. (B)

    The interior of the lipid bilayer is nonpolar and does not allow polar molecules to easily pass through, thus the need for a transport protein.

24. (B)

    The opening statement is a hint about the hypothesis. The presence of digestive enzymes in the intestinal tissue and blood stream is thought to be the cause of shock. How does physical trauma contribute to the presence of digestive enzymes in tissue and blood? This leads to the hypothesis: The drop in blood pressure and subsequent reduced blood flow causes the breakdown of the protective mucosal lining of the small intestine (B). (A) and (C) are incorrect because the experimental description does not assess the involvement of the immune system response or the build up of waste products in the small intestine. It is possibly true that digestion will stop or slow down after a physical trauma, but there are probably sufficient reserves in the form of glycogen to maintain internal energy levels and continue producing ATP (D).

25. **(A)**

Digestive enzymes have the ability to break down all the classes of organic molecules, and will if they come into contact with them. Part of the compartmentalization of the digestive system means that enzymes released into the lumen of the small intestine stay there and carry out tissue destruction only there. In fact, most digestive enzymes are produced in an inactive form because they have the potential to digest the cells that produce them, so (A) is correct. (B) is not true. The digestive enzymes would not be recognized as non-self and so are not antigenic. (C) is not true because enzymes can be produced repetitively. In the normal course of digestion, many enzymes that are produced are released with waste material and must be replaced. (D) is not true. The conditions outside the small intestine are not sufficiently different in temperature, pH, or other conditions to denature the enzymes. It would be a good thing to have them denature, anyway, because then they could not act on the host's tissues.

26. **(B)**

The enzyme blockers may work by competitively inhibiting the active site of the digestive enzymes. If this is the way they work, the inhibitor has to be complementary to the chemistry of the active site and must out-compete the normal substrate for the enzyme. So, a high concentration of the blocking molecules needs to be present and the molecules would have to be polar or ionic. A nonpolar molecule would not interact with an ionic molecule. They would not be complementary to the active site.

27. **(C)**

The change in mass over the experimental period is 50% for the plants kept in the light. The number of days, however, is four (not five—a common mistake made by students). So 50%/4 days = 12.5%/day, which rounds to 13%.

28. **(B)**

Potassium ions pumped into the guard cells make the guard cells hypertonic and water follows the $K^+$ ions and causes the guard cells to swell and open the stomata. If there are too few $K^+$ ions, stomata will remain closed and water loss will drop. So, the most significant drop in loss of mass would occur in the plants that are losing the most water—those in the light.

29. **(A)**

The loss of mass in the plants is an indirect measure of the loss of water in each group of plants. Therefore, the greatest loss of water occurred in the plants exposed to constant light. Water is lost when it moves from relatively high water potential to relatively low water potential so the greatest loss of water occurs where the water potential is lowest.

30. (B)

Competitive inhibition involves blocking the active site of an enzyme with a molecule that is not the usual substrate for the enzyme. The inhibitor would have to be chemically and/or physically similar to the usual substrate as indicated in this question. There is no evidence in the description of this process that the enzyme is denatured (A) or that there is an alternative enzymatic pathway (D). (C) is incorrect because an allosteric inhibitor would have a shape that is suited for the allosteric site and would not be similar to the usual substrate.

31. (C)

The trioses that are formed from photosynthesis make up the increased mass, but the mass of the trioses is mostly due to the mass of the carbon and oxygen of $CO_2$ that has been reduced, not the mass of the hydrogens with their electrons that were oxidized from water (A). The minerals incorporated into plant molecules contribute only a minute amount of the mass of the tree as they are used in only trace amounts.

32. (B)

It is important to remember that operons, the sequential genes that control a metabolic pathway in bacteria, are either repressible or inducible and that the repressor is a separate protein produced by a regulatory gene. If the operon is inducible, the substrate (lactose in this case) binds with the regulatory protein changing its conformation and making it inactive, i.e., unable to bind with the operator site on the gene. Thus, RNA polymerase is able to transcribe the genes of the operon. Lactose does not interact with DNA (C), and the presence of lactose allows transcription so (A) is incorrect and ribosomes would normally interact with mRNA, not proteins free in the cytoplasm (D).

33. (B)

The interaction of CAP/DNA directly stimulates gene expression, that is, a change in a variable increased the change it induces, the description of positive feedback. (A) is incorrect because it is not a negative feedback (the opposite condition); the response was not independent of its stimulus (C); and the on/off switch for the *lac* operon was just the lactose (D). The CAP activator was a rate changer.

34. (A)

ATP is produced by the flow of hydrogen from a higher concentration to a lower concentration through ATP synthase. The establishment of the hydrogen gradient occurs by the transfer of electrons along the electron transfer chain. If cyanide stops the electron transfer, no gradient will be established and ATP production will cease.

## 35. (A)

Cyanide disrupts the action on an enzyme involved in the electron transport chain. This points to its function as a coenzyme. Many coenzymes act as electron donors and acceptors in concert with enzymes. Phosphorylation of ADP does not occur until the end of the electron transport chain (B). Oxidation of glucose (C) may slow and formation of $CO_2$ (D) may cease, but only as an indirect result of stopping the electron movement along the electron transport chain.

## 36. (B)

The oxidation of food molecules ultimately sends electrons to the electron transport chain. The final electron acceptor is oxygen which, combined with hydrogen ions, forms water. This is the major source of metabolic water. Some water is generated through condensation reactions, but much of this is consumed by hydrolysis reactions.

## 37. (B)

Conserved processes are ancient and usually indicate fundamental, vital processes. (A) and (D) are both correct statements about glycolysis, but can be said about other processes and do not necessarily make a statement about how fundamental glycolysis is. (C) is an incorrect statement. Glycolysis only generates 2 ATP per molecule of glucose.

## 38. (C)

The plants show three different flowering responses to the differing light applications. Some plants respond to a long period of light and shorter period of dark. Some respond to a short period of light and longer period of dark. Some plants do not seem to be affected by the change in periods of dark and light.

## 39. (B)

The poinsettias received 9 hours of light in treatment 2 and had a 98% flowering rate. However, when the plants received 9 hours of light in treatment 4, the flowering rate dropped to 7%. The length of dark time was broken up and the poinsettias did not receive 15 hours of continuous dark. Potatoes (A) flowered when exposed to 15 hours of light and no other stretch of light, and tomatoes (C) and roses (D) did not change their flowering percentages under any of the differing treatments.

## 40. (C)

Potatoes are called long-day plants; however, careful examination of the data shows that potatoes successfully flower with only 9 hrs of continuous daylight. Potatoes did successfully flower when their night length was varied such that the period of continuous darkness was less than 9 hours. This means that the potatoes are more influenced by the length of darkness than the length of light.

41. **(C)**

    The establishment of an acquired immune response relies on MHC antigen presentations. In the case of cancer, the cancer cells have evaded the monitoring by cells of the immune system. The virus targeted the cancer cells and destroyed them providing cancer cell fragments. These fragments would include some of the faulty cancer-causing proteins and, once presented, would generate an acquired immune response tailored to the tumor cells. (A) and (D) are wrong because each one implies that a virus is cellular. An inflammatory response (B) is not part of the acquired immune response.

42. **(C)**

    (A), (B), and (D) all are major contributors to the success of the immune system, however, what all of these depend upon is the vast diversity and specificity of protein conformations. There must be a vast array to complement to the many potential antigens that attack an organism.

43. **(A)**

    The increased risk of a particular disorder in a group of organisms indicates the increased presence of a gene or genes. This would occur as a result of elective breeding, which decreases genetic diversity. Selective breeding (B) does not increase the rate of mutations, and as for (C), there is no evidence that these dogs are exposed to environments that increase the likelihood of mutations. (D) would imply that cancer is "catching" (contagious) and a genetic disorder cannot be caught like a cold.

44. **(D)**

    Substrate-level phosphorylation yields a net of two ATP in glycolysis and two in the citric acid cycle. However, oxidative phosphorylation yields about 32 ATP.

45. **(A)**

    RRaa (red, axial flower) × RrAA

    RR × Rr yields 50% RR (red) and 50% Rr (pink).

    aa (axial) × AA (pure breeding terminal) = 100% Terminal

    Therefore, 50% of the offspring will be RrAa, pink, with terminal flowers, and 50% will be RRAa, red, with terminal flowers.

46. **(B)**

Cells need to maximize their surface area to volume ratio in order to maximize the exchange of materials across the cell membrane.

47. **(C)**

Quorum sensing is a response to the density of microorganisms. If the density is optimum, the bacteria can maximize their success by sending signals that will increase their ability to form a biofilm, release toxins, and exchange DNA.

48. **(B)**

It is advantageous for a large mammal to be in a cooler climate as it will neither overheat nor cool off too quickly. Conversely, a small mammal can cool off more quickly and would have a selective advantage in a warmer climate.

49. **(D)**

Diffusion (A) is a passive process and occurs with any liquid or gas (B) such as when the scent of an onion diffuses across a room to be smelled. (C) is describing a cell in a solution that is probably hypertonic to the cell and will cause the cell to shrink due to water loss. If the concentration of the salt is exactly the same as the solute concentration in the cell, it is possible that the cell would be isotonic, but that is probably not the case. With isotonic solutions (D), there is no net movement of water or solutes across the membrane.

50. **(D)**

Conjugation (A) is the exchange of genetic material between bacteria through a sex pilus. Transformation (B) is bacterial uptake of genetic material from the environment. Transduction (C) is the inadvertent transfer of genetic material from one bacteria to another by a bacteriophage. Antibiotics (D) do the opposite of genetic variation; they exert selective pressure for a particular phenotype, decreasing variation.

51. **(B)**

The reduction of carbon dioxide occurs during the light independent reactions of photosynthesis. (A), (C), and (D) are all necessary preparatory steps to provide the raw materials and energy to reduce carbon dioxide.

52. **(A)**

   While the citric acid cycle does not use oxygen, it will not take place without oxygen, creating the "pull" on electrons. (B) and (D) occur after the citric acid cycle during oxidative phosphorylation, and (C) is an activity of photosynthesis.

53. **(C)**

   Glycolysis takes place with or without oxygen present. However, if $NAD^+$ is not present to act as an electron acceptor, glycolysis will stop and the supply of ATP will stop also.

54. **(D)**

   Mitochondria and plastids have a double lipid bilayer consistent with having been a lipid bilayer bound organism engulfed by a larger lipid bilayer organism. (A), (B), and (C) are all true and evidence for eukaryotic evolution.

55. **(A)**

   The maternal contribution to an egg includes mRNA and proteins that are asymmetrically introduced into the egg in an inactive form. Once the egg is fertilized and cell division begins, the mRNA and proteins will be unevenly distributed within the cells. Once activated, these maternal mRNA and proteins will interact with the DNA of the particular cells they are in, causing different cells of the embryo to produce different proteins, and this determines the basic body plan for the organism. (B) is incorrect because the chromosomes are not distributed asymmetrically. They reside in the nucleus. (C) and (D) are correct statements, but occur after the asymmetric distribution of maternal contributions.

56. **(C)**

   In order to prepare a plasmid vector, a common restriction endonuclease must be used on the source DNA and on the plasmid. Then the source DNA is mixed with the plasmid and ligase to seal the DNA and the plasmid. Then the bacteria are plated on a medium to select for transformants.

57. **(B)**

   The graph depicts an endergonic reaction in which the reactants are lower in energy than the products. (B), synthesis of a dipeptide, is the only endergonic reaction in the choices.

58. **(D)**

   Enzymes cannot make a reaction happen that would not happen on its own. They can only lower the energy of activation ($E_a$) so that the reaction can take place under conditions that could occur.

59. **(D)**

   The action potential of a neuron relies on establishing an electrochemical gradient (charge/polarity) across the neuron's membrane. This is done by pumping three sodium ions out of the cytoplasm for every two potassium ions pumped into the cell. There is a net positive charge outside the cell relative to the inside.

60. **(A)**

   The specificity of proteins is responsible in order to recognize the substrate with which it will interact. There are some actions that do not require energy (B), there are some enzyme reactions that do not rely on membrane proteins (C), and (D) is simply an incorrect statement. Membrane potential does involve proteins that are specific, but not ones that function as described in the question.

61. **(B)**

   A depolarization event triggers the opening of voltage-gated ion channels. This change in polarity triggers a large change in the membrane voltage generating an action potential. (A) Polarizing the membrane sets the neuron up for an action potential and brings the membrane back to resting potential. (C) ATP is necessary for the establishment of the membrane potential, but not the depolarization. (D) A minimum supply of the neurotransmitter acetylcholine is necessary to propagate the action potential to the next nerve cell in the series.

62. **(C)**

   If a farm has been abandoned, there is soil present and this represents secondary succession. The pioneer organisms would be grasses.

63. **(A)**

   Cyclin-dependent kinases are necessary to carry out mitosis. This action would stop mitosis. However, DNA replicates during interphase, so it will have replicated. Incorrect or faulty DNA may contribute to aborting mitosis. The cell will not disintegrate, though it may under apoptosis, and it will not enter the $G_0$ phase if it is a faulty cell.

# Detailed Explanations of Answers

## Section I, Part B

1. **20.74**

    There were 1,000 offspring in this cross (572 + 428). The null hypothesis states that half were expected to be tall and half short; thus the expected values are $0.5 \times 1000 = 500$. The chi-square formula will be given on the formula sheet and is: $\chi^2 = \Sigma(o-e)^2/e$. These values are then used in a table like the one below:

    | Class | Expected | Observed | o−e | (o−e)² | (o−e)²/e |
    |---|---|---|---|---|---|
    | Tall | 500 | 572 | 572 − 500 = 72 | (72)² = 5184 | 5184/500 = 10.37 |
    | Short | 500 | 428 | 428 − 500 = −72 | (−72)² = 5184 | 5184/500 = 10.37 |

    $$\chi^2 = 10.37 + 10.37$$

    $$\chi^2 = 20.74$$

2. **$1.37 \times 10^8$ cal/24 hours ($1.37 \times 10^5$ C/day)**

    $\Delta$ heat = (blood flow rate)(time)($\Delta$ temp)(blood specific heat)

    Solve the problem in the following way:

    $$\frac{5L}{min} \times \frac{60\,min}{hr} \times \frac{24\,hr}{d} \times \frac{1000\,c}{1L^\circ C} \times 19^\circ C = 1.37 \times 10^8 \text{ cal}/24 \text{ hours } \left(1.37 \times 10^5 \text{ C}/24 \text{ hours}\right)$$

3. **1.2**

    The rate of this chemical reaction is calculated as the change in the product over time ($\Delta Y/\Delta X$). At one minute, there is 1.4 mL of product and at four minutes there is 3.8 mL of product. Thus, 3.8 ml − 1.4 ml/3 min − 1 min = 1.2 mL/min.

4. **0.4**

The Hardy-Weinberg equation will be provided on the formula sheet and is:

$$p^2 + 2pq + q^2 = 1 \text{ and } p + q = 1.$$

Because the population is in Hardy-Weinberg equilibrium, you can assume that the frequency of each allele in the population today would be similar to that collected in the year 2000. Light fur is the dominant allele and both homozygous dominant individuals ($p^2$) and heterozygote individuals ($2pq$) will have the light phenotype. Thus, $0.84 = p^2 + 2pq$.

Essentially, you are solving for $q$, so

$$q^2 = 1 - (p^2 + 2pq)$$

$$q^2 = 1 - 0.84 = 0.16$$

$$q = \text{the square root of } 0.16 = 0.4$$

5. **55 g/m²**

The squirrels receive 110 g/m² from the trees, lose 45 g/m² to the atmosphere from metabolic processes, and lose 10 g/m² to predation from the coyotes.

$$110 - 45 - 10 = 55 \text{ g/m}^2$$

6. **3/32**

The genotype of the offspring is A_bbcc, with the _ representing either AA or Aa. The odds of producing each offspring are:

$$A\_ = \frac{3}{4}$$
$$bb = \frac{1}{2}$$
$$cc = \frac{1}{4}$$
$$cc = \frac{1}{4}$$

So $= \dfrac{3}{4} \times \dfrac{1}{2} \times \dfrac{1}{4} = \dfrac{3}{32}$

# Detailed Explanations of Answers

## Section II

1.

(*Note:* 10-point maximum—this rubric is not specific because there is a wide variety of answers that are possible. The answer must remain related to the specific plant and animal systems chosen throughout the response. Although a table is never an acceptable response format, the table below demonstrates how a response should flow and how each part relates to the other parts. Each answer needs detailed elaboration.)

|  | (a) Identification; system affected | (b) Describe recognition | (c) Explain Response | (d) Evolutionary Advantage |
|---|---|---|---|---|
| Animal | Neurotransmitter-nervous system | Specific post-synaptic membrane receptors | Muscle contraction to avoid a predator | Move/avoid danger/survive |
| Plant | Night length-seeds/germination | Photoreceptors | Change in shape of phytochrome transduces signal that causes cell proliferation, leading to growth of plant embryo | Coordinates germination to optimal environmental conditions |

## Possible Answers and Explanations:

(a) 2-point maximum

*1 point for identification of a plant signal and the system—both must be answered correctly to earn the point.*

- Possible examples include: plant hormones; amount of light, day or night length; direction of light; or description of any other plant tropism.

*1 point for identification of an animal signal and the system—both must be answered correctly to earn the point*

- Possible examples include: peptide or steroid hormones of the endocrine system; neurotransmitters of the nervous system; hormones controlling gamete formation in the reproductive system; antigen recognition by the immune system; and growth factors.

(b) 2-point maximum—response must be related to the answer in part (a) to receive credit

*1 point for the explanation of recognition in the plant cell*

*1 point for the explanation of recognition in the animal cell*

- Possible answers for both the plant and animal include: plasma membrane-bound receptors, cytoplasmic receptors, and nuclear receptors.

- Plant specific receptors include: phytochromes, photoreceptors, mechanical receptors.

(c) 4-point maximum—up to 2 points for any of the following statements:

- Receptors are specific for one ligand.

- Receptors change shape when their ligand binds.

- Shape change can activate the receptor and cause it to interact with other molecules, causing transduction of the signal.

- Important to homeostasis because many diseases are caused by malfunctioning receptors.

*1 point for plant explanation and 1 point for animal explanation; answers must be specific for the process described in parts (a) and (b).*

- Possible answers for both plant and animal responses include: death of the cell, cell proliferation, gene regulation, and increased or decreased activity of an enzyme.

- Animal only responses include: transduction of a nervous signal, contraction of muscle, activation of an endocrine gland, and antibody production.

(d) 2-point maximum—1 point for each of the following statements:

- Efficient recognition provides a phenotypic benefit that makes the organism more likely to survive.

- The appropriate response provides a benefit that makes the organism more likely to survive.

- Coordination of responses increases opportunities for reproduction (when to flower, mating behavior).

2. (10-point maximum)

   (a) 3 points

   - *1 point for labeled axes and title*
   - *1 point for correctly graphed data points*
   - *1 point for stating population 2 had the higher rate of transpiration*

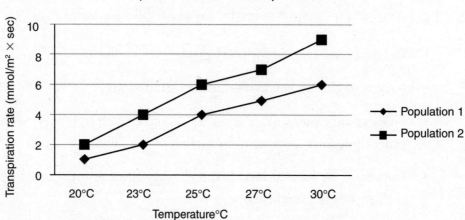

   (b) 3-point maximum—one point for any of the following statements:

   - The explanation should pinpoint a difference in the environment between the two populations.

- Population 1 lives where plants will be shaded by other plants and where there is high humidity. Both of these situations would have high water potential and decrease rates of transpiration.

- Population 2 lives in an area that would be sunny and is windy. Both of these conditions decrease water potential, making water loss through transpiration higher.

- Definition of water potential.

- Explanation of water potential—water moves from high to low water potential.

(c) 2-point maximum—only the first two answers are graded. One point for any of the following statements:

1. Plants can cross pollinate.

2. Plants produce viable offspring.

3. Plants produce fertile offspring.

4. Comparison of chromosome number.

(d) 2-point maximum—explanations must be associated with the exact piece of data provided in part (c). One point for any of the following statements – note the statements are linked to those in part (c).

1. Definition of species.

2. Definition of species.

3. Definition of species.

4. Same number of chromosomes or high similarity suggests same species.

3. 4-point maximum. See Chapter 11 for the discussion of countercurrent exchange.

(a) 2 points can be earned

- *1 point for correct description of countercurrent exchange:* In countercurrent exchange, vessels or channels move in opposite directions, and materials can flow between the two tubes.

- *1 point for correctly explaining a specific system.* For example: In thermoregulation, vessels containing warm and cool blood flow next to one but another in opposite directions; heat is exchanged between the vessels, warming the cool blood. In gills and fish, the blood flow in the gills moves in the opposite direction to the flow of water, maximizing the $O_2/CO_2$ gradient.

    Other examples could include: osmoregulation in a mammalian kidney.

(b) Two points can be earned.

- *1 point for the correct definition of homeostasis*

- *1 point for the correct application of homeostasis to the situation chosen in part a.* For example, an endothermic animal living in cooler temperatures must maintain its core temperature. As warm blood flows from the core to the extremities, it passes by cool blood returning to the core. Some of the heat will flow from the warm blood to the cool blood, bringing its temperature closer to core temperature. This means the animal will not have to expend as much energy to maintain its core temperature.

4. 4-point maximum

    *2 points for analysis of data. One point can be earned for each statement.*

    - The results of generation 2 occur because a cross between homozygous dominant (tan) and homozygous recessive (brown) will result in all heterozygous (tan) offspring. In this case, all 29 offspring are tan.

    - Offspring in generations 3-7 consistently occur in nearly a 3:1 ratio.

    - Calculations of $p$ and $q$ in generations 3-7 should be approximately $p = 0.5$ and $q = 0.5$.

    *2 points for the explanation of H-W equilibrium*

    - The population is in H-W equilibrium because generations 3 to 7 maintain almost constant allele frequencies of tan and brown.

A population in H-W equilibrium exhibits the following characteristics (correct answer must list at least 3): large population, no mutations, random mating, no natural selection, and no gene flow.

5. 4-point maximum

    (a) 3 points can be earned, one point for each of the following statements:

    - Explanation of the function of at least three enzymes used (primase, ligase, helicase, polymerase, a topoisomerase).

    - DNA polymerase can only add to the 3′ end of DNA, thus synthesis always occurs in the 5′ to 3′ direction.

    - Replication of the leading strand moves continuously and into the replication fork, while replication of the lagging strand is interrupted and moves away from the replication form.

    - Leading strand uses one RNA primer; lagging strand uses multiple RNA primers

    - Okazaki fragments are generated along the lagging strand because DNA polymerase adds to the 3′ end only. Fragments are sealed with ligase.

    - Nucleotide sequence is maintained by complementary base pairing (A-T and C-G).

    - Mistakes occurring in either strand are repaired by nuclease

    (b) 1 point can be earned for one of these three statements:

    - The same four nucleotides (A, T, C, and G) are found in all living things.

    - Many of the enzymes used in prokaryotic replication are the same as in eukaryotic (polymerase, helicase, primase).

    - Commonality of code—same amino acids for codon sequence, ability to introduce human genes into bacteria and obtain human protein—recombination.

6. 4-point maximum

    (a) 2 points can be earned; *one point for each of the following. Only one point is awarded for the correct discussion of an example. If more than one example is given, only the first is scored.*

    - Negative feedback mechanism decreases the initial signal.

    - Possible examples include: hypothalamic regulation of internal temperature in mammals; glucagon or insulin regulation of glucose levels; *lac* or *trp* operon in prokaryotes; secretion of bicarbonate by the pancreas; and production of testosterone by the testes.

    - Negative feedback mechanisms maintain homeostasis by keeping an organism in balance. If more or a molecule is needed, it is produced until levels return to normal.

    (b) 2 points can be earned; *one point for each of the following. Only one point is awarded for the correct discussion of an example. If more than one example is given, only the first is scored.*

    - Positive feedback mechanisms increase or amplify the response.

    - Possible examples include: oxytocin to increase labor, oxytocin control of lactation, and blood platelet accumulation.

7. 3-point maximum

    (a) 1 point for explanations of either of the following:

    - Sickle cell anemia shows heterozygote advantage.

    - Having some sickle cells prevents malarial infection and is maintained in populations living in areas with high prevalence of malaria.

    (b) 2-point maximum for any of the following statements:

    - Wobble substitution at the 3$^{rd}$ nucleotide often doesn't matter because there is a redundancy in the genetic code.

- If the substituted amino acid has similar chemical properties (acidic for acidic or hydrophobic for hydrophobic), it may not affect the shape of the protein.

- If it's an enzyme and the substituted amino acid is not in the active site, the active site may still bind its substrate.

8. 2-point maximum; 1 point is earned for any of the following statements:

    - The inner membrane of the mitochondria is folded many times into cristae.

    - Embedded in the folds are ATP synthase, the enzyme responsible for ATP production.

    - Increased surface area enhances the synthesis of ATP and thus increases the efficiency of cellular respiration.

    - The electron transport chain occurs in the membrane, so more surface area provides space for many ETCs.

# Answer Sheet

## Part A

1. Ⓐ Ⓑ Ⓒ Ⓓ
2. Ⓐ Ⓑ Ⓒ Ⓓ
3. Ⓐ Ⓑ Ⓒ Ⓓ
4. Ⓐ Ⓑ Ⓒ Ⓓ
5. Ⓐ Ⓑ Ⓒ Ⓓ
6. Ⓐ Ⓑ Ⓒ Ⓓ
7. Ⓐ Ⓑ Ⓒ Ⓓ
8. Ⓐ Ⓑ Ⓒ Ⓓ
9. Ⓐ Ⓑ Ⓒ Ⓓ
10. Ⓐ Ⓑ Ⓒ Ⓓ
11. Ⓐ Ⓑ Ⓒ Ⓓ
12. Ⓐ Ⓑ Ⓒ Ⓓ
13. Ⓐ Ⓑ Ⓒ Ⓓ
14. Ⓐ Ⓑ Ⓒ Ⓓ
15. Ⓐ Ⓑ Ⓒ Ⓓ
16. Ⓐ Ⓑ Ⓒ Ⓓ
17. Ⓐ Ⓑ Ⓒ Ⓓ
18. Ⓐ Ⓑ Ⓒ Ⓓ
19. Ⓐ Ⓑ Ⓒ Ⓓ
20. Ⓐ Ⓑ Ⓒ Ⓓ
21. Ⓐ Ⓑ Ⓒ Ⓓ
22. Ⓐ Ⓑ Ⓒ Ⓓ
23. Ⓐ Ⓑ Ⓒ Ⓓ
24. Ⓐ Ⓑ Ⓒ Ⓓ
25. Ⓐ Ⓑ Ⓒ Ⓓ
26. Ⓐ Ⓑ Ⓒ Ⓓ
27. Ⓐ Ⓑ Ⓒ Ⓓ
28. Ⓐ Ⓑ Ⓒ Ⓓ
29. Ⓐ Ⓑ Ⓒ Ⓓ
30. Ⓐ Ⓑ Ⓒ Ⓓ
31. Ⓐ Ⓑ Ⓒ Ⓓ
32. Ⓐ Ⓑ Ⓒ Ⓓ
33. Ⓐ Ⓑ Ⓒ Ⓓ
34. Ⓐ Ⓑ Ⓒ Ⓓ
35. Ⓐ Ⓑ Ⓒ Ⓓ
36. Ⓐ Ⓑ Ⓒ Ⓓ
37. Ⓐ Ⓑ Ⓒ Ⓓ
38. Ⓐ Ⓑ Ⓒ Ⓓ
39. Ⓐ Ⓑ Ⓒ Ⓓ
40. Ⓐ Ⓑ Ⓒ Ⓓ
41. Ⓐ Ⓑ Ⓒ Ⓓ
42. Ⓐ Ⓑ Ⓒ Ⓓ
43. Ⓐ Ⓑ Ⓒ Ⓓ
44. Ⓐ Ⓑ Ⓒ Ⓓ
45. Ⓐ Ⓑ Ⓒ Ⓓ
46. Ⓐ Ⓑ Ⓒ Ⓓ
47. Ⓐ Ⓑ Ⓒ Ⓓ
48. Ⓐ Ⓑ Ⓒ Ⓓ
49. Ⓐ Ⓑ Ⓒ Ⓓ
50. Ⓐ Ⓑ Ⓒ Ⓓ
51. Ⓐ Ⓑ Ⓒ Ⓓ
52. Ⓐ Ⓑ Ⓒ Ⓓ
53. Ⓐ Ⓑ Ⓒ Ⓓ
54. Ⓐ Ⓑ Ⓒ Ⓓ
55. Ⓐ Ⓑ Ⓒ Ⓓ
56. Ⓐ Ⓑ Ⓒ Ⓓ
57. Ⓐ Ⓑ Ⓒ Ⓓ
58. Ⓐ Ⓑ Ⓒ Ⓓ
59. Ⓐ Ⓑ Ⓒ Ⓓ
60. Ⓐ Ⓑ Ⓒ Ⓓ
61. Ⓐ Ⓑ Ⓒ Ⓓ
62. Ⓐ Ⓑ Ⓒ Ⓓ
63. Ⓐ Ⓑ Ⓒ Ⓓ

# Part B

FREE-RESPONSE ANSWER SHEET

For the free-response section, write your answers on sheets of blank paper.

# Glossary

**abiotic factors** refer to the nonliving parts of an ecosystem such as water and sunlight.

**acetyl CoA** pyruvate is converted to this by adding coenzyme A; it will go into the Krebs cycle.

**acquired immunity** the immune response of vertebrate animals that utilizes B cells and T cells to target pathogens.

**action potential** the electric signal that travels down a neuron.

**action spectrum** the indicator of the wavelengths of light that can drive photosynthesis.

**activation energy** the energy required to start a chemical reaction.

**active site** the location on an enzyme where substrate binds and where the chemical reaction occurs.

**active transport** moving a substance against its concentration gradient; it requires energy and a transport protein.

**adaptation** a trait inherited by an organism that increases its chance for survival.

**adenosine triphosphate** see ATP.

**adhesion** one substance sticks to another; usually refers to water; for example water sticks to xylem in plants, but can also mean the interactions between cells that contributes to an organism's form.

**aerobic respiration** catabolic process that converts glucose into ATP using oxygen; occurs in the cytosol and the mitochondrion.

**alcohol fermentation** follows glycolysis; reduces NADH to NAD$^+$, making $CO_2$ and ethanol.

**allele** one form of a gene that gives a phenotypic effect.

**allosteric regulation** one molecule binds to a protein and affects its ability to bind other molecules at a separate site.

**alpha helix** in a protein, the coiled region of amino acids resulting from hydrogen bonds between individual amino acids; the secondary structure of proteins.

**alternation of generations** in plants and algae; the life cycle alternates between a diploid form (sporophyte) and a haploid form (gametophyte).

**amino acid** the building block of proteins; consists of an amine group, a carboxyl group, and a variable R group.

**amphipathic** often refers to phospholipids or transmembrane proteins embedded in the phospholipid bilayer; molecule with a hydrophobic region and a hydrophilic region.

**anabolic pathway** uses energy to build complex molecules from smaller molecules.

**anaerobic respiration** an ATP generating pathway that uses an electron transport chain where oxygen is not the final electron acceptor.

**anaphase** phase of mitosis where the sister chromatids have separated and are moving to the poles of the cell.

**aneuploidy** either having extra chromosomes or missing chromosomes; triploidy and monoploidy are examples.

**antibody** a protein secreted by a B cell that binds to specific antigens.

**antigen** a molecule that binds to either a B or T cell and triggers an immune response.

**apoptosis** programmed cell death.

**aquaporin** protein channels that facilitate the movement of water into a cell.

**atom** a chemical unit that cannot be broken down. It is composed of protons, neutrons, and electrons.

**atomic mass** the mass of an atom; the sum of the protons and neutrons. This can change because atoms of a particular element can have different numbers of neutrons, therefore different masses.

**atomic number** the number of protons in an atom; this is the means of identification of an atom because all atoms of a particular element have the same number of protons.

**ATP** adenosine triphosphate; a nucleotide with three phosphate groups attached. ATP couples exergonic reactions with endergonic reactions by the transfer of the third phosphate group.

**autotroph** an organism that converts simple molecules into organic molecules, usually using energy from the sun; also called a producer.

**bacteriophages** viruses that attack bacteria.

**behavior** in animals a physiological action that occurs in response to an external stimulus.

**bile** a chemical produced by the liver and stored and released from the gall bladder into the small intestine; it emulsifies fat to aid in digestion.

**biotechnology** the process of applying molecular research to the production of useful products.

**biotic factors** the living components of an ecosystem.

**blastula** the hollow ball of cells that results from the initial set of cell divisions that occurs after fertilization.

**carrying capacity** the number of individuals of a population that an ecosystem can support based upon the resources available; it is not a constant number but fluctuates with changes to the ecosystem.

**catabolic pathway** metabolic pathway that hydrolyzes large molecules into smaller ones.

**cDNA** the DNA that results from the use of reverse transcriptase to make DNA from mRNA. The cDNA will not have introns.

**cell-mediated immune response** the type of acquired immunity that uses T cells to destroy infected cells.

**cellular respiration** the process of converting organic molecules, usually carbohydrates, into ATP; includes both anaerobic and aerobic processes.

**chemiosmosis** the phosphorylation of ADP to ATP that occurs as a result of the establishment of a proton motive force.

**chemiosmotic photophosphorylation** the phosphorylation of ADP to ATP that occurs as a result of the establishment of a proton motive force using the electrons excited by light energy.

**chemoautotrophs** an organism that produces chemical energy using carbon dioxide and inorganic molecules as an energy source.

**chemoreceptors** a molecule, cell, or organ receptive to a particular chemical signal.

**chlorophyll a** the green pigment that absorbs that harnesses the light energy of photosynthesis.

**chloroplast** the organelle in eukaryotic photoautotrophs that carries out photosynthesis.

**cholesterol** a lipid composed of four fused rings; an important component of the cell membrane and a precursor to many steroids.

**chromatin** the diffuse form of chromosomes that is apparent during interphase of the life cycle.

**chromosome** DNA bearing the genetic instructions. In prokaryotes, there is one circular chromosome. In eukaryotes, the chromosomes are a linear arrangement that is often associated with proteins.

**cleavage** the process of cell division that occurs in embryonic development; forms a hollow ball of cells called a blastula.

**cloning** the use of biotechnology to make an exact copy of a gene or a whole organism.

**codon** the sequence of three nucleotides in DNA or RNA that codes for an amino acid or stop.

**cofactor** an organic or inorganic substance necessary for the proper functioning of an enzyme.

**commensalism** a symbiotic relationship in which one organism is helped and the other is neither helped nor harmed.

**communication** the sending, receiving, and interpreting of signals between living organisms.

**community** all populations living and interacting together in a given area.

**competition** two organisms in the same ecosystem try to obtain and use the same resources; within the same species is intraspecific; between two different species is interspecific.

**consumers** organisms that rely on other organisms (usually plants or algae) for their energy resources; so named because they must "consume" other organisms.

**contractile vacuole** a specialized organelle used to remove excess water; typically found in aquatic protists and algae.

**convergent evolution** process in which organisms that are not closely related evolve similar traits as a result of having been subjected to similar selective pressures.

**countercurrent exchange** the exchange of heat or substances between two fluids moving in opposite directions; used for thermoregulation and gas exchange in many animals.

**crossing-over** gene exchange between non-sister chromatids that occurs during prophase I of meiosis. It increases genetic variation and separates linked genes.

**cuticle** in plants, the waxy, outer layer of stems and leaves.

**cyanobacteria** bacteria that photosynthesize releasing oxygen into the atmosphere.

**cyclic electron flow** the process in which excited electrons in photosystem I are not accepted by an electron acceptor; instead, they return to the photosystem reaction center. The process generates ATP but does not reduce $NADP^+$.

**cytokine** a protein secreted by a cell that assists in triggering an immune response.

**cytoplasmic determinants** maternal contributions of mRNA and proteins to the egg prior to fertilization; these play a major regulatory role in body plan formation.

**cytotoxic T cell** a cell in the immune system that when activated kills cells infected with pathogens.

**dehydration synthesis** the formation of larger molecules from smaller ones by formation and removal of a molecule of water.

**deletion** a mutation that occurs when one or more nucleotide bases are removed from a sequence of DNA.

**denatured** a change in a molecule that causes the molecule to lose its functional shape.

**derived trait** a trait that is present in an organism that was absent from the last common ancestor of the taxonomic group being considered.

**differentiation** in animal development the specialization of a cell for a certain function.

**diffusion** the movement of particles from an area of high concentration to an area of low concentration.

**dihybrid** involves two separately inherited traits or an individual who is heterozygous for a trait.

**diploid** a cell with two complete sets of chromosomes, one set from each parent.

**disturbance** in ecology, some event that causes a change in an ecosystem by either removing organisms or changing the distribution of natural resources; examples include hurricanes and introduced species.

**divergent evolution** process in which organisms that are related with underlying similarities evolve very different traits as a result of having been subjected to very different selective pressures.

**domain** a part of a protein that folds and functions independently of another part of the protein.

**dominant gene** a gene that masks the presence of another gene.

**duplication** a replication of a segment of DNA.

**ectotherm** an animal whose body temperature is determined by its external environment (contrast with endotherm); also termed poikilotherm or cold-blooded.

**electron transfer chain** a series of protein complexes in the thylakioid membrane of the chloroplast or the cristae of the mitochondria. These complexes alternate between oxidized and reduced states to pass high energy electrons down an energy gradient harnessing the energy to establish a proton motive force.

**electronegativity** a chemical property of an atom or element that describes its tendency to pull electrons toward itself.

**element** a substance composed of only one kind of atom.

**endergonic** a chemical reaction that consumes energy (positive change in free energy).

**endocytosis** the uptake into a cell of macromolecules or particles; occurs by an infolding of the cell membrane that ultimately surrounds the material with a vesicle.

**endomembrane system** the membranes within a cell, as well as the plasma membrane, that allow for compartmentalizing reactions. The Golgi apparatus, the nuclear envelope, vesicles, vacuoles, and lysosomes are components of the endomembrane system.

**endoplasmic reticulum** part of the endomembrane system of a cell that is responsible for formation and modification of proteins as well as for membrane formation and synthesis of lipids. It may be rough, associated with ribosomes, or smooth, not associated with ribosomes.

**endotherm** an animal that keeps warm by using heat it generates during cellular respiration; contrast with ectotherm. Also termed homeotherm or warm-blooded.

**entropy** a measure of the disorder in a system.

**enzyme** biological catalysts which lower activation energy, thus increasing the rate of a chemical reaction; typically proteins.

**epigenetics** inherited changes that do not result from a change to DNA.

**epiglottis** a flap of cartilage that folds over the trachea during swallowing to prevent food from entering the trachea.

**epitop** the region on an antigen that is recognized and bound by an antibody.

**equlibrium** the state of a system in which there are no net changes occurring; a system in this state cannot do work.

**eutrophication** "well nourished"; occurs when an aquatic ecosystem receives an abundance of nutrients (typically nitrogen and phosphorus).

**exergonic** a chemical reaction that releases energy (negative change in free energy).

**exponential growth** growth of a population that is not limited; shows a J-shaped curve when graphed over time.

**exocytosis** the release of macromolecules such as proteins from the cell that occurs when a vesicle merges with the cell membrane.

**exon** a segment of DNA or RNA that codes for part of a polypeptide. Exons are spliced together.

**extant** still in existence, not extinct.

**F1/F2** the letter designation for the first and second generation of any genetic cross under consideration.

**facilitated diffusion** The passive movement of polar or ionic solutes into or out of a cell through a membrane protein.

**fermentation** An anaerobic process that starts with glycolysis to generate a small amount of ATP. Reactions at the end of glycolysis regenerate $NAD^+$ so that glycolysis can continue. Alcohol or lactic acid are the usual products of this regeneration process.

**fluid-mosaic model** The model of the cell membrane that reflects its mixed composition of phospholipids, proteins, glycoproteins, and other molecules as well as the motions of these materials within the membrane.

**food chain** a representation showing the path of energy transfer from trophic level to trophic level.

**frameshift mutation** A change in the DNA sequence due to an addition or a deletion of DNA. The result will be a series of incorrectly placed amino acids.

**free energy** The amount of energy in a system that is available to do work.

**gametes** a reproductive cell with a haploid number of chromosomes; usually an egg or a sperm.

**gastrulation** The movement of embryonic cells to establish the three tissue layers, the endoderm, mesoderm, and ectoderm.

**gel electrophoresis** A technique for separating fragments of a molecule by size using an electrical charge. Often used to separate DNA fragments that have been cut with endonucleases.

**gene** unit of inherited information coded for by a particular sequence of DNA.

**gene flow** movement of genes between populations via fertile individuals.

**gene pool** all of the alleles for all of the loci within a population.

**genetic engineering** manipulation of DNA usually to put together DNA from two different sources.

**germination** in plants, the process in which the embryonic plant grows from the seed.

**glycolysis** a process that occurs in all organisms to produce ATP; involves the oxidation of glucose and production of pyruvate.

**Golgi apparatus** an organelle that is part of the endomembrane system. It receives, stores, and modifies proteins destined for the cell membrane or for secretion.

**granum** a stack of thylakoid membranes within a chloroplast.

**gross primary productivity** the amount of sunlight energy that is converted into chemical energy, measured in biomass, that producers in an ecosystem synthesize.

**guard cells** in vascular plants, the two cells on either size of a stoma that control its opening and closing.

**haploid** a cell, typically a gamete cell, that has one unpaired set of chromosomes present.

**helicase** an enzyme that catalyzes the unwinding of DNA during replication.

**helper T cell** a lymphocyte in the immune system that, when activated, triggers B cells and cytotoxic T cells.

**hemoglobin** the protein in red blood cells responsible for oxygen transport.

**hepatic portal vessel** a large vein that transports all the products of digestion to the liver.

**heterotroph** an organism that must eat other organisms to gain energy; sometimes called consumers.

**heterozygote** a genotype with two different alleles for the gene in question.

**hibernation** an animal response to cold in which core temperature and metabolism drop in an attempt to conserve energy.

**histones** small proteins associated with DNA; they are usually positively charged; hence, the negative DNA remains wound around them unless acted on to release their hold.

**homeobox** sequences of DNA found as part of homeotic genes, that are highly conserved and act as gene regulators during embryonic development.

**homeostasis** the ability of an organism to maintain a stable internal environment, despite changes in its surroundings; for example, a human maintains a fairly constant temperature of 37 °C.

**homeotic** a gene that plays a major part in the formation of a body plan.

**homologous pair** a pair of chromosomes, one from each parent, that carry information for the same set of genes.

**homozygote** a genotype with two of the same alleles for a gene in question. The genotype may be homozygote dominant or recessive.

**humoral response** the immune system defense that utilizes B cells and antibodies; found in body fluids.

**hypertonic** a solution with a relatively higher concentration of solutes in comparison to another solution. It would have a lower water potential.

**hypothalamus** a region in the mammalian brain that serves as a thermostat by activating physiological processes that either warm or cool the animal.

**hypotonic** a solution with a relatively lower concentration of solutes in comparison to another solution. It would have a higher water potential.

**immunoglobulin** proteins (antibodies) secreted by B-cell lymphocytes in response to the presence of a pathogen. Immunoglobulins are classified by how they interact with pathogens.

**induced fit** the change in an enzyme's shape because of the substrates present in its active site.

**induction** a change in a group of cells due to regulatory signals received from another set of cells.

**innate behaviors** in animals, genetically determined and developmentally fixed responses to an environmental stimulus.

**innate immunity** the fast-acting defense mechanism in animals and plants that is not specific for a certain pathogen.

**introns** areas of DNA that do not code for a protein but are transcribed and removed after transcription and before translation.

**isomers** molecules that have the same molecular formula but a different geometric arrangement of the constituent atoms and therefore different chemical properties.

**isotonic** a solution with the same solute concentration in comparison to another solution.

**isotopes** atoms of a particular element that differ in the number of neutrons present. This affects the atom's mass but not its identity or behavior.

**kinesis** a behavior in which an animal moves randomly in response to an environmental stimulus; for example, a pill bug exposed to light moves more often, but in no specific direction.

**kinetic energy** the energy of motion.

**lacteal** lymphatic vessels in the small intestine that absorb and transport the products of fat digestion.

**lagging strand** the strand of DNA that is replicated discontinuously from the 5′ to 3′ direction, away from the replication fork.

**law of independent assortment** a principle that allele pairs on two nonhomologous chromosomes will sort independently of each other.

**law of segregation** a principle that a homologous pair of chromosomes will separate and go to different gametes during meiosis.

**leading strand** the strand of DNA that is replicated continuously in the 5′ to 3′ direction toward the replication fork.

**learned behaviors** in animals, variable responses to an external stimulus that depend on past experiences of an organism.

**ligase** the enzyme responsible for sealing the growing strand of DNA with phosphodiester bonds.

**logistic growth** growth of a population that is limited by factors such as resource availability; when graphed over time, it shows an S-shaped curve and the population levels off at carrying capacity.

**long-day plants** in order to produce flowers, these plants must be exposed to light periods that are longer than the critical period.

**lumen** the open space inside the intestines or the endoplasmic reticulum or any other tubular structure.

**lymphocytes** small white blood cells that participate in the immune response.

**lysis** the rupture of a cell membrane.

**meiosis** the replication, splitting, and distribution of chromosomes within the gametes of an organism. It requires that the number of chromosomes within the individual be halved but ensures that one member of each pair of chromosomes is distributed to each gamete.

**mesophyll** the inner layer of tissue in a leaf that is composed of photosynthetic cells and vascular tissue.

**metabolism** the sum total of chemical reactions that go on in a cell.

**microRNAs** small fragments of single-stranded RNA that bind to complementary fragments of mRNA, thus preventing translation of a protein.

**microvillus** minute hairlike structures that project from a cell increasing its surface area.

**migration** the movement of animals over a long distance due to environmental cues.

**mitosis** the replication of chromosomes and distribution to two new nuclei resulting in two daughter cells with the same genetic composition as the original cell; the kind of cell division that occurs in somatic cells.

**monoclonal antibodies** antibodies produced by a single line of cloned cells such that the antibodies are all identical.

**monohybrid cross** a cross between two organisms in which only one gene locus is considered.

**monosaccharide** the simplest carbohydrates; the monomer unit of disaccharides and polysaccharides.

**morphogenesis** the development of an organism according to the overall body plan.

**mutation** often considered the raw material of variation, it is a change in the DNA of an individual.

**mutualism** a symbiotic relationship in which both organisms benefit.

**NAD⁺** a coenzyme that acts as an electron acceptor; the primary electron acceptor in cellular respiration.

**NADP⁺** a coenzyme that acts as an electron acceptor in photosynthesis.

**negative feedback** a common biological regulatory process in which a stimulus triggers a response that counteracts the stimulus.

**net primary productivity** the amount of chemical energy made by producers in an ecosystem (GPP) minus the amount of chemical energy needed for cellular respiration; NPP = GPP − respiration.

**noncyclic electron flow** the one-way electron flow from water to NADP⁺ in the light dependent reactions of photosynthesis that yields NADPH and ATP to drive the light independent reactions of photosynthesis.

**nonpolar covalent bond** a bond between atoms in which the electrons are shared equally.

**nucleosomes** the complex of DNA and histone proteins that makes up the structural unit of a chromosome.

**nucleotides** the monomer unit of DNA or RNA that consists of a nitrogenous base, a sugar, and a phosphate group.

**okazaki fragments** the short segments of DNA that are formed from the discontinuous replication of the lagging strand.

**operon** found in prokaryotes; consists of a cluster of genes whose products function in a metabolic pathway, under the control of a promoter and an operator.

**origin of replication** the area on DNA that is recognized as the starting point for DNA replication.

**osmoregulation** the ability of organisms to balance water gain and loss as well as maintain appropriate concentrations of solutes within their cells.

**osmosis** the diffusion of water across a selectively permeable membrane.

**oxidation-reduction** chemical reactions that result in a shift of electrons; one reactant will donate one or more electrons (the reducer) and be oxidized and one reactant will gain the electrons (the oxidizer) and become reduced.

**pancreas** an organ of the body that contributes to glucose homeostasis and produces digestive enzymes for all classes of organic molecules.

**parasitism** a type of symbiotic relationship in which one organism benefits at the expense of the other organism; an example is a tick sucking the blood of a dog.

**parental** the original organisms considered in a genetic cross.

**passive transport** movement of solutes or water from high concentration to low concentration without using energy.

**pathogen** a disease-causing microorganism that can infect another organism.

**pepsin** the active form of a proteinase secreted in the stomach.

**peristalsis** involuntary wavelike constriction of muscles of the esophagus and intestines.

**phagocytosis** engulfing extracellular material by extension of the cell membrane around the material to form a vesicle.

**phenotype** the observable physical, physiological, or biochemical expression of a gene within an organism.

**phospholipids** a lipid molecule that is the main component of the plasma membrane; it is a heterogeneous type of molecule with a polar and nonpolar region.

**phosphorylation** the transfer of a phosphate group, usually from ATP, to a reactant in an endergonic reaction.

**photons** a discrete bundle of light energy; it is the energy that will boost the electrons in photosynthesis.

**photoperiodism** in plants, a physiological response to the amount of light received; the length of day can affect flowering, seed germination, and production of leaves.

**photosystems** a set of pigment and protein complexes in photoautotrophs that capture light energy to drive the redox reactions of photosynthesis.

**phototropism** the growth response of a plant towards or away from a light stimulus.

**phytochrome** a light receptor in plants that regulates many plant activities.

**pigment** a substance that absorbs particular light energy and reflects or transmits other light energy.

**pinocytosis** a form of endocytosis that takes in extracellular liquids.

**pioneer species** in ecological succession, the first organisms to move in to a rocky area; typically moss and lichen.

**plasmid** an extrachromosomal, circular piece of DNA found in prokaryotes that replicates separately from the chromosome.

**plasmolysis** the shriveling of the cytoplasm of a cell that has a wall. The loss of water causes the cell to shrivel and the membrane to pull away from the wall.

**point mutation** a single change to DNA that results in a change in the placement of an amino acid.

**polar covalent bond** a bond between atoms in which the electrons are not shared equally.

**polymerase** an enzyme that catalyzes the synthesis of DNA or RNA.

**polymerase chain reaction (PCR)** a use of biotechnology to make many copies of a length of DNA.

**population** a group of organisms of the same species that can freely interbreed and produce fertile offspring.

**population density** the number of individuals of a species in a given area; populations with higher densities put great stresses on the resources of their ecosystem.

**positive feedback** a biological regulatory process that amplifies a response to a stimulus.

**potential energy** energy based on position; in biology, it is usually stored bond energy.

**predator** an organism that hunts and kills another organism (the prey) for food.

**prey** an organism that is killed by a predator.

**primase** an enzyme that catalyzes the placement of an RNA primer on DNA to enable DNA synthesis.

**producers** the organisms in an ecosystem that synthesize organic molecules from carbon dioxide, typically through photosynthesis.

**proto-oncogenes** a normal gene within a cell that, if transformed, can become an oncogene, potentially a cancer-causing gene.

**recessive** the term applied to a gene that is masked by the presence of another gene in the heterozygous genotype. The phenotype is only apparent if both the alleles are the same.

**recombinant DNA** DNA from two nonhomologous sources put together.

**recombinases** enzymes that have the ability to rearrange DNA bringing different segments of DNA into close proximity; they function in the DNA rearrangements that contribute to the huge variety of B- and T- cell receptors.

**restriction endonucleases** enzymes capable of recognizing short stretches of DNA. These enzymes originated in bacteria as a defense against bacteriophages.

**reverse transcriptase** an enzyme capable of catalyzing the synthesis of a cDNA from mRNA.

**ribosome** organelle responsible for reading and translating mRNA into a polypeptide.

**ribozyme** an RNA molecule that can act as a catalyst; the name comes from **ribo**nucleic acid + en**zyme** because it displays properties of both molecules.

**saturated fatty acid** the hydrocarbon tail has only single bonds between the carbons of the chain; these fatty acids are very nonpolar and will pack together decreasing cell membrane fluidity.

**short-day plants** in order to produce flowers, these plants must have periods of dark longer than the critical period.

**short tandem repeats** DNA segments in noncoding regions of DNA that are used as genetic markers.

**signal peptide** the initial sequence of a peptide which signals it for escort to the rough endoplasmic reticulum.

**species** There are several ways to look at species but the most common definition is: A group of organisms that freely interbreed and produce fertile offspring. However, species distinction has to do with presence/absence of gene flow between populations.

**species biodiversity** the number and type of species found in an ecosystem; higher biodiversity indicates a more stable ecosystem.

**sphincters** circular muscles at various junctures along the digestive tract that control movement of material from one compartment of the digestive tract to another.

**stomata (plural)** the pores on the leaf epidermis that allow for gas exchange between the inside of the leaf and the air; each stoma (singular) is surrounded by two guard cells which open or close the stoma.

**stroma** the clear, fluid-filled space within the chloroplast; the site of the light independent reactions of photosynthesis.

**substrate** the particular molecule(s) that enter the active site of a protein and are acted on by the protein.

**substrate-level phosphorylation** the addition of a phosphate group from an activated organic molecule to ADP to form ATP.

**succession** the step-by-step series of changes in species found in an ecosystem after it is disrupted; divided into primary and secondary succession.

**symbiosis (symbiotic relationships)** an interacting, long-term relationship between two different species that live closely together; examples include parasitism, commensalism, and mutualism.

**taxis** an animal behavior in which the animal moves towards (positive) or away from (negative) an environmental stimulus; for example, an animal moving towards light exhibits positive phototaxis.

**taxonomy** the science of classifying and naming organisms.

**telomeres** the ends of linear chromosomes that shorten with each round of replication; they prevent shortening of coding DNA.

**thermolabile** an enzyme whose reaction is slower or nonfunctional at even a small increase in ideal temperature.

**thermophilic** organisms that tolerate high temperatures.

**thermoregulation** maintaining a fairly constant internal body temperature.

**thylakoid membrane** the green, chlorophyll-laden membrane within the chloroplast; this is the site of the light-dependent reactions.

**thylakoid space** the area bounded by the thylakoid membrane; the site of the proton motive force that drives chemiosmosis.

**topoisomerase** an enzyme that moves ahead of the replication form and prevents kinking by cutting the DNA and allowing it to swivel.

**totipotent** a cell that has the ability to give rise to any other type of cell in the organism.

**transcription** the synthesis of mRNA from a DNA template.

**transcription factors** protein that binds to an enhancer or promoter and helps bind RNA polymerase to initiate transcription.

**transformation/transformants** the modification of DNA through the use of biotechnology or the modification of a cell from a normal cell to a cancerous cell. A transformant is a cell that has undergone transformation.

**translation** the synthesis of a protein from an mRNA transcript.

**translocation** relocation of a chromosomal segment usually to a different chromosome.

**transpiration** the loss of water from a plant through evaporation.

**triplet** the three nucleotide pairs that make up a codon that codes for a particular amino acid.

**trophic level** the position in a food chain an organism occupies.

**tropism** a growth response in plants either towards or away from a stimulus; growth towards is positive and growth away is negative.

**tumor suppressor genes** a gene that produces a protein that inhibits cell division, thus inhibiting the potentially uncontrolled growth of a tumor.

**turgor pressure** occurs in plant cells in a hypotonic solution; as water fills up the cell, it pushes the cell membrane against the cell wall, keeping the plant from wilting.

**unsaturated fatty acid** the hydrocarbon tail has one or more double bonds between the carbons of the chain. These fatty acids have slight bends at the site of the double bond and this inhibits close packing and contributes to cell membrane fluidity.

**vacuole** a large membrane-bound vesicle.

**vectors** the plasmid used to carry a DNA segment into a host cell.

**vesicle** a sac made of membranous material.

**villus** fingerlike projections of the small intestine that increase the surface area for digestion and absorption.

**water potential** the potential energy of water, determined by solute potential and pressure potential; water moves from areas of high water potential to areas of low water potential.

**xylem** the vascular tissue in plants that transports water, minerals, and ions that are absorbed by the roots from the soil; consists of hollow bundles of cells that act like tubes throughout the plant.

**zygote** A fertilized egg with its diploid chromosome number restored.

# Appendix

Appendix

# AP Biology Equations and Formulas

## Statistical Analysis and Probability

**Standard Error**

$$SE_{\bar{x}} = \frac{s}{\sqrt{n}}$$

**Mean**

$$\bar{x} = \frac{1}{n}\sum_{i=1}^{n} x_i$$

**Standard Deviation**

$$S = \sqrt{\frac{\sum(x_i - \bar{x})^2}{n-1}}$$

**Chi-Square**

$$\chi^2 = \sum \frac{(o-e)^2}{e}$$

### Chi-Square Table

| | Degrees of Freedom | | | | | | | |
|---|---|---|---|---|---|---|---|---|
| p | 1 | 2 | 3 | 4 | 5 | 6 | 7 | 8 |
| 0.05 | 3.84 | 5.99 | 7.82 | 9.49 | 11.07 | 12.59 | 14.07 | 15.51 |
| 0.01 | 6.64 | 9.32 | 11.34 | 13.28 | 15.09 | 16.81 | 18.48 | 20.09 |

### Laws of Probability

If A and B are mutually exclusive, then P (A or B) = P (A) + P (B)

If A and B are independent, then P (A and B) = P (A) × P (B)

### Hardy-Weinberg Equations

$p^2 + 2pq + q^2 = 1$     $p$ = frequency of the dominant allele in a population

$p + q = 1$     $q$ = frequency of the recessive allele in a population

$s$ = sample standard deviation (i.e., the sample based estimate of the standard deviation of the population)

$\bar{x}$ = mean

$n$ = size of the sample

$o$ = observed individuals with observed genotype

$e$ = expected individuals with observed genotype

Degree of freedom equals the number of distinct possible outcomes minus one.

### Metric Prefixes

| Factor | Prefix | Symbol |
|---|---|---|
| $10^9$ | giga | G |
| $10^6$ | mega | M |
| $10^3$ | kilo | k |
| $10^{-2}$ | centi | c |
| $10^{-3}$ | milli | m |
| $10^{-6}$ | micro | µ |
| $10^{-9}$ | nano | n |
| $10^{-12}$ | pico | p |

Mode = value that occurs most frequently in a data set

Median = middle value that separates the greater and lesser halves of a data set

Mean = sum of all data points divided by number of data points

Range = value obtained by subtracting the smallest observation (sample minimum) from the greatest (sample maximum)

## Rate and Growth

**Rate**
$dY/dt$

**Population Growth**
$dN/dt = B - D$

**Exponential Growth**
$$\frac{dN}{dt} = r_{max} N$$

**Logistic Growth**
$$\frac{dN}{dt} = r_{max} N \left( \frac{K - N}{K} \right)$$

**Temperature Coefficient $Q_{10}$**
$$Q_{10} = \left( \frac{k_2}{k_1} \right)^{\frac{10}{t_2 - t_1}}$$

**Primary Productivity Calculation**
mg $O_2$/L × 0.698 = mL $O_2$/L

mL $O_2$/L × 0.536 = mg carbon fixed/L

## Surface Area and Volume

**Volume of Sphere**
$V = 4/3 \pi r^3$

**Volume of a cube (or square column)**
$V = l\,w\,h$

**Volume of a column**
$V = \pi r^2 h$

**Surface area of a sphere**
$A = 4 \pi r^2$

**Surface area of a cube**
$A = 6a$

**Surface area of a rectangular solid**
$A = \Sigma$(surface area of each side)

---

$dY$ = amount of change
$t$ = time
$B$ = birth rate
$D$ = death rate
$N$ = population size
$K$ = carrying capacity
$r_{max}$ = maximum per capita growth rate of population

$t_2$ = higher temperature
$t_1$ = lower temperature
$k_2$ = metabolic rate at $t_2$
$k_1$ = metabolic rate at $t_1$
$Q_{10}$ = the *factor* by which the reaction rate increases when the temperature is raised by ten degrees

$r$ = radius
$l$ = length
$h$ = height
$w$ = width
$A$ = surface area
$V$ = volume
$\Sigma$ = Sum of all
$a$ = surface area of one side of the cube

---

## Water Potential ($\Psi$)

$\Psi = \Psi_p + \Psi_s$

$\Psi_p$ = pressure potential

$\Psi_s$ = solute potential

The water potential will be equal to the solute potential of a solution in an open container, since the pressure potential of the solution in an open container is zero.

**The Solute Potential of the Solution**
$\Psi_s = -iCRT$

$i$ = ionization constant (For sucrose this is 1.0 because sucrose does not ionize in water)

$C$ = molar concentration

$R$ = pressure constant ($R$ = 0.0831 liter bars/mole K)

$T$ = temperature in Kelvin (273 + °C)

## Dilution - used to create a dilute solution from a concentrated stock solution

$C_i V_i = C_f V_f$

$i$ = initial (starting)      $C$ = concentration of solute

$f$ = final (desired)      $V$ = volume of solution

## Gibbs Free Energy

$\Delta G = \Delta H - T \Delta S$

$\Delta G$ = change in Gibbs free energy

$\Delta S$ = change in entropy

$\Delta H$ = change in enthalpy

$T$ = absolute temperature (in Kelvin)

**pH** $= -\log[H^+]$

# Index

## A

Abiotic factors, 266
Absorption spectra of photosynthetic pigments, 89
Acetyl-CoA, 95
Acidic solution, 78
Action spectrum, 89
Activated hormone-receptor complexes, 213
Activation energy of a reaction, 85
Active sites, 246
Active transport, 113–115
Adaptations, 31
Adaptive radiation, 40, 56
Adenine nucleotide, 161
Adenosine diphosphate (ADP), 69
Adenosine triphosphate (ATP), 71, 90, 138
   breakdown of, 69
   in catabolic reactions, 68
   chemical structure of, 69
   production, 47
   synthase molecules, 97
Aerobic respiration, 93, 151
Agarose, 196
Age structure diagram, 238–239
Alcoholic fermentation, 98–99
Algae, 223
Allele, 33
Allopatric speciation, 40
Allosteric activation and inhibition, 249
Allosteric regulator, 249
Alpha glucose, 81
Alpha subunit, 262
Alternation of generations, 135
Altruistic behavior, 218
Amino acids, 83
Anabolic reactions, 68
Anaerobic fermentation, 94
Anaerobic prokaryotes, 55
Anaerobic respiration, 151
Analogous structures, 46
Anaphase, 176
Anaphase I, 178
Aneuploidy, 198
Animal cell, 102
Animal communication, 218
Animal development, 138–141
Antibody, 261–262
AP Biology exam
   free-response section, 8
   multiple-choice section, 7–8
   scoring process for, 9
   strategies for attempting multiple-choice section, 10–11
   types of questions, 11
AP Biology labs, 8–9
Apical ectodermal ridge, 140
Apoptosis, 141
Aquaporins, 110, 143
*Arabidopsis thaliana,* 125
Archaea, 123
Artificial selection, 39, 59
   data collection and analysis, 60–61
   sampling, 60
   statistical methods, 60
Asexual reproduction, 70, 137
Atomic mass of an atom, 75
Atomic number of an element, 75
Autopolyploidism, 180
Autotrophic organisms, 119, 149
Autotrophs, 73
Avery, Oswald, 158

## B

Bacterial transformation with cDNA, 196
Base pairing, 160
Basic solution, 78
B cells, 261–262
Beta glucose, 81
Binary fusion, 137
Binomial nomenclature, 51
Biofilm, 254
Biogeography, 45
Biological chemistry, 74–76
Biological molecules, 80–85
Biological processes, 69
Biological systems, 78
Biological systems, source of energy for, 71, 73
Biomass, 271
Biosynthesis, 74
Biosynthesis of polymers, 79
Biotechnology, 194
Biotic factors, 266
BLAST (Basic Local Alignment Search Tool)
   data analysis using, 65
   scientific practices, 64–65
Blastula, 139
Bottleneck effect, 36
Bound ribosome, 168
Breathing, 126
Buddings, 137

## C

*C. elegans,* 141
Calcium ions ($Ca^{2+}$), 215
Calvin-Benson cycle, 90–92
Calvin cycle, 73, 88
Cancer cells, 229
*Canis lupus,* 51
Carbohydrates, 80–81
Carbon fixation, 91
3-carbon molecules, 91
5-carbon molecules, 91
Carotenoids, 89
Carrier proteins, 110
Carrying capacity, 237

Catabolic reactions, 68
Catalyzed reaction, 246
CDNA, 195
Cell cycle, 173–177
Cell differentiation, 124–125, 140, 193
Cell membrane structure, 106–109
Cell membrane transport
 active, 113–115
 passive, 109–112
Cell respiration, equation for, 93
 parts of, 94–97
Cells
 components of, 123
 of multicellular eukaryotic organisms, 123–124
 response to relative solution concentration, 112
Cell theory, 123
Cell-to-cell communication, 211–212
Cell-to-cell signaling proteins, 124
Cellular respiration, 72, 84, 93–94, 104, 126, 151, 153, 254, 274
Cellular structures, 47
Cellulose, 254
Cell walls, 109
Channel protein, 110
Chargaff, Erwin, 159
Chargaff's rules, 159–160
Chase, Martha, 158
Chemical signaling, 213
Chemiosmosis, 93
Chemiosmotic photophosphorylation (chemiosmosis), 90
Chemoreceptors, 126
Chi-square formula, 228
Chlorohyll, 88
Chlorophyll b, 89
Chloroplast, 87, 88, 105
Cholecystokinin, 254
Cholesterol, 82, 107
Chromatids, 176, 230
Chromosomal abnormalities, 229
Chromosomes, 158
Cladograms, 49–50, 64

Cleavage, 139
Cloning DNA, 195
*Clostridium botulinum*, 200
Clownfish, 223
Coacervates, 54
Codon, 56, 147
Codon AUG codes, 56
Coenzyme A (CoA), 95
Coenzymes, 84, 86, 248
Cofactors, 86, 248
Cognition, 219–221
Commensalism, 255
Communities, 237–239, 255
Comparative anatomy, 45
Comparative embryology, 46
Competitive inhibitors, 248–249
Condensation reactions, 79
Conserved structures and processes of genes, 47
Consumers, 73, 239, 272
Cooperative behaviors, 223
Coupled reaction, 71
Covalent bonds, 76
*Crash Course for AP Biology*, 3
Cretaceous extinction, 56
Cretaceous period, 56
Cristae, 104
Crossing over, 179, 229
Cross-pollination, 223
Cyclic AMP (cAMP), 215
Cyclins, 176
Cytokinesis, 174, 176
Cytosol, 106

### D

Darwin, Charles, 29, 41
Daughter cell, 137
Deciduous trees, 132
Deforestation, 257
Denaturing, 83
Density-dependent factors, 238
Density-independent factors, 238
Derived traits, 50
Differentiation, 139
Diffusion, 109–110, 143
 science practices, 145–149

Digestive system, 251–252
 food movement and digestive activities, 253
Diploid, 174
Diploid organisms, 63
Direct contact signaling, 213
Directional selection, 38, 59
Diversifying selection, 38
Diversity of ecosystems, 266–267
DNA
 elongation, 163
 as genetic material, 157–159
 helicases, 163
 initiation, 163
 lagging strand, 164
 leading strand, 164
 ligase, 164
 polymerase, 163
 primase, 163
 replication, 163–165
 structure of, 159–162
 *vs* RNA, 165–166
DNA replication, mistakes in, 199
DNA sequence comparison, 64–65
DNA viruses, 200, 229
Down syndrome, 180, 230
Dynamic equilibrium, 111

### E

Earth's ecosystems, 242
Earthworms, 130
EcoRI, 194
Ecosystem, distribution of, 257–258
Ectothermy, 71
E-flashcards, 3
Eldredge, Nile, 41
Electronegative partner, 76
Electrons of an atom, 75
Electron transport, 72
Electron transport chain (ETC), 89–90, 96–97
Embedded proteins, 107
Endergonic reactions, 68–69
Endocrine signaling, 213
Endocytosis, 115

Endomembrane system of a eukaryotic cell, 103–106
Endoplasmic reticulum (ER), 121, 168
　route to protein, 122
Endoplasmic reticulum (ER) of organelles, 104
Endothermic, 128
Endothermy, 70–71
Energy, defined, 67
Energy capture, 72–73
Energy coupling, 71–72
Energy currency, 74
Energy cycles, 241
Energy storage and growth, 70
Energy transfers, 76
Enhancer, 192
Enhancer region, 192
Enterogastrone, 254
Entropy, 67
Environmental catastrophes, 257
Enzymatic reactions, 248
　science practices, 281–286
Enzyme pathways, 127
Enzyme reaction cycle, 247
Enzymes, 85–86, 246
Enzyme-substrate complex, 247
Epigenetics, 192, 197, 264
Ethanol, 98
Eubacteria, 123
Eukaryotes, 19, 46, 48, 55–56, 87–88, 93, 101–102, 109, 163, 166, 174, 189–190, 196
　gene regulation in, 191–194
Eukaryotic cells, 48, 55, 87, 101
Eukaryotic DNA, 195
Eukaryotic sexual reproduction, 199
Evolutionary history, models of, 49–51
Evolutionary relatedness, 64
Evolution of bacteria, 48
Exergonic reactions, 68–69
Exocytosis, 115
Exons, 166
Exponential growth, 237
External factors, 197
Extracellular digestion, 252

## F

Facilitated diffusion, 110
Fallopian tubes, 139
Fatty acids, 81
Feedback mechanisms, 126–129
Fermentation, 73, 94, 98–99
Fertilization, 138
　of an egg, 139
Flatworms, 130
Flow of energy in an ecosystem, 239–240
Fluid mosaic model of plasma membrane, 107
Food chain, 239, 272
　science practices, 272–274
Food web, 240
Fossils, 45
Founder effect, 36
Fragmentation, 137
Frameshift mutation, 170
Franklin, Rosalind, 159
Free energy, 67
Free-response section, 8
　achieving success in, 20–24
　pointers, 27
　sample response in, 24–26
　strategies for attempting, 18–20
Free ribosomes, 168
Fruit fly behavior analysis, 278–281
*FSH* (follicle stimulating hormone), 138
Full-length practice test, 3
Fungal cell walls, 109
Fungi, 223

## G

Galápagos Islands, 30
Gametophyte, 135
Gamma subunit, 262
Gastrin, 254
Gel electrophoresis, 196
Gene flow, 37
Gene frequencies of a population, 34
Gene pool, 33
Gene regulation, 189
Genetic diversity, 265
Genetic drift, 36, 265
Genetic engineering
　applications, 194–195, 197
　techniques, 195–197
Genetic modification, 39–40
Genetic mutation, 137
Genital opening alignment, 40
Genotype, environmental effects on, 264
Geographic isolation, 41
Geological events, 258
Germination, 136
Ghrelin, 254
Gibberellins, 136
Glucose ($C_6H_{12}O_6$), 80–81
Glycolysis, 47, 73, 80, 93–95, 98, 248
Golgi apparatus, 105, 121, 123
Gould, Stephen Jay, 41
G-proteins, 215
Gradualism, 30, 41
Grana, 87
Griffith, Frederick, 158
Gross primary productivity (GPP), 271
Growth factors, 238

## H

Haldane, John, 53
Haploid, 174
Hardy-Weinberg equation, 265
Hardy-Weinberg Equilibrium, 34, 36–37, 62
　example problem for, 34–35
HeLa cells, 229
Helicases, 163
Henrietta Lacks (HeLa cells), 199
Hermaphroditism, 137
Hershey, Alfred, 158
Heterotrophs, 73
Heterozygosity, 263
Heterozygote advantage, 198
Hibernation, 132
Hierarchical classification system, 51

Histones, 174, 191
Homeostasis, 119, 123
　disruptions in, 131–132
Homeostatic mechanisms
　continuity and, 130
　　that change over time,
　　130–131
Homeotic genes, 49, 140, 193
Homologous pair, 174, 178
Homologous structures, 45
Horizontal gene transfer, 199
*Hox* genes, 49, 140
Human papillomavirus (HPV),
　229
Human population, estimate
　of, 242
Hydrogen bonding, 77
Hydrogen pump, 114
Hydrolysis reactions, 79
Hydrophilic (water loving)
　head, 107
Hydrophilic (water loving)
　substances, 78
Hydrophobic (water fearing)
　fatty acid, 107
Hydrophobic (water loving)
　substances, 78
Hypertonic solution, 111
Hypothalamus, 128
Hypothesized relationships, 51
Hypotonic solution, 111

# I

Immunoglobulin, 262
Independent assortment, 230
Induced fit, 247
Inducible operon, 190
Inductive signaling, 140
Information exchange, 217–218
Innate behaviors, 222
Inorganic phosphate, 69
Insulin levels and feedback
　mechanism, 129
Integral proteins, 108
Internal factors, 197
Internal organelles, 103
Interspecific competition, 256
Intracellular digestion, 252

Intracellular receptors, 214
Introns, 166
Ionic bonds, 75–76
Isotonic solution, 110

# K

Karyotypes, 229
Keystone species, 267
Kinesis, 278
Krebs (citric acid) cycle, 73, 93,
　95

# L

Lactic acid fermentation, 98–99
Lagging strand, 164
Lamarck, Jean-Baptiste, 29
Law of thermodynamics, 67
Leading strand, 164
Leaf loss in plants, 136
Learned behaviors, 222
Lectins, 227–228
Leptin, 254
LH (luteinizing hormone), 138
Life-history strategy, 70
Ligase, 164, 195
Light-dependent reaction,
　88–91
Linnaeus, Carolus, 51
Lipids, 81–82
Liposomes, 54
Logistic growth, 237
Long-distance signaling, 213
Lyell, Charles, 30
Lymphocytes, 261–262
Lyse, 111
Lysogenic cycle, 199–200
Lysosomes, 105
Lytic cycle, 199

# M

MacLeod, Colin, 158
Macromolecules, 79, 119
Malthus, Thomas, 30
Mammalian thermoregulation,
　128
Mathematical modeling
　concepts, 62
　science practices, 63
Matrix, 104
Matter, 74
McCarty, Maclyn, 158
Medulla, 126
Meiosis, 137, 177–180
　modeling, 229–230
Meiosis II, 177–178
Membrane-bound receptors,
　214
Membrane proteins, function
　of, 108
Menstrual cycle, 137
Menstrual phase, 138
Mesodermal cells, 140
Mesophyll, 92
Metanephridia, 130
Metaphase, 175
Methylation, 264
　of DNA, 191
Microevolution, 36–37
MicroRNAs (miRNAs), 141
Microspheres, 54
Microtubules, 178
Migration, 223
Miller, Stanley L., 54
Mini-tests, 3
Missense mutation, 170
Mitochondrion of organelles,
　104
Mitosis
　effects of environment on,
　　227–228
　modeling, 227
　phases of, 176
Molecular biology, 46
Molecular interactions within a
　living system, 245–249
Molecular variation, 261–263
Monomers, 79
Monosaccharides, 80
Monosomic chromosome, 198
Morphogenesis, 124, 193
MPF (mitosis promoting
　factor), 177
MRNA, 121
MRNA transcript, 166
Multicellular organism, 47

Multiple-choice section, 7–8
    avoiding common errors, 14
    grid-in questions, 15–18
    predicting right answers, 13–14
    process of elimination, 12
    strategies for attempting, 10–11
Mutation, 37, 46, 197–198, 200
Mutualism, 255
Mutualistic symbiotic relationships, 223

### N
NAD+, 248
NADH, 94–96
NADP+, 89, 248
NADPH per mole, 90
Na+/K+ pump, 113–114
Natural selection, 29–30, 37
    postulates, 31
Natural theology, 29
Negative control, 189
Negative feedback, 190
Negative feedback loops, 127–128
Nephrons within kidneys, 130
Net primary productivity (NPP), 239, 272–273
Nitrogenous bases, 158, 160
Nondisjunction, 180
Nonpolar bonds, 76
Nonrandom rating, 37
Nonsense mutation, 170
Nucleic acids, 83–84
Nucleolus of organelles, 104
Nucleosomes, 191
Nucleotides, 84
Nucleus of an atom, 75
Nucleus of organelles, 103–104

### O
Oogenesis, 137
Oparin, Alexander, 53
Oparin-Haldane hypothesis, 53–54
Operons, 127, 190

Organelles, 119–121, 251
Organic interactions within a living system, 250–255
Organic nutrients, 53
Organismal processes, 46–48
Organs, communication between, 125–126
Organ systems, 119
Origin of life on Earth, 55–56
Osmosis, 110–111, 143
    science practices, 145–149
Ovarian cycle, 138
Oxaloacetate (OAA), 95
Oxidative phosphorylation, 72
Oxpecker birds, 223
Oxytocin, 129

### P
*Pachycara brachycephalum,* 263
Paracrine signaling, 213
*Paramecium,* 112
Parasitism, 255
Parthenogenesis, 137
Passing of characteristics to offspring, 53
Passive transport, 109–112
Peripheral proteins, 108
Peristalsis, 252
Peroxisomes, 106
Phagocytosis, 115
Phenotype, 33, 59, 197
Phospholipids, 82, 85, 261
    bilayer structure of the plasma membrane of cells, 47
Phosphorylation, 71
Phosphorylation cascade, 214
Photoperiodism, 131–132
Photorespiration, 92
Photosynthesis, 72, 80, 87–88, 132, 274
    environmental factors influencing, 87
    light reactions of, 90–91
    science practices, 149–151
    steps, 88–92
Photosystems, 89
Phylogenetic trees, 49
Phylogeny, 49

Physical contact signaling, 213
Phytochromes, 125
Pinocytosis, 115
Plant cell walls, 109
Plants, response to adverse environmental conditions, 136–137
Plasma membrane, 106
Plasma membrane of prokaryotes, 123
Plasmids, 190, 194
Point mutations, 170
Polar covalent bonds, 76
Polymerase chain reaction (PCR), 196
Polypeptides, 125
Polyploidism, 180
Polyploidy, 198
Population density, 237
Population dynamics, 265
Population genetics, 33
Population interactions, 255–257
Positive control, 189
Positive feedback, 129, 190
Post-transcriptional control, 192
Postzygotic barriers, 40
Potential energy, 70
Predator–prey relationships, 255
Pressure potential, 112, 144
Prey, 255
Prezygotic barriers, 40
Primase, 163
Primitive Earth, 54
Programmed cell death, 141
Prokaryotes, 46, 48, 55–56, 87, 101–102, 123, 125, 137, 166, 189, 199, 251
    gene regulation in, 190–191
Prokaryotic cells, 48, 55, 101, 123
Proliferative phase, 138
Promoter sequence, 166
Prophage DNA, 200
Prophase, 175
Prophase I, 177
Protein domain changes, 246
Protein kinases, 214–215
Protobionts, 54

Protonephridia, 130
Proto-oncogenes, 194
Provirus, 200
Pr/Pfr phytochrome, 125
Punctuated equilibrium, 41
Pyrimidines, 158
Pyruvate, 95

## Q

Quaternary structure, 83
Quorum sensing, 211

## R

Random assortment, 179
Random fertilization, 179
REA AP All Access system, 1–2
REA Study Center, 2
 suggested 8-week ap study plan, 4
Recombinant DNA, 195–196
Recombinases, 262
REDOX reactions, 76
Reduction–oxidation reactions. see REDOX reactions
Regeneration, 91
Repressible operon, 190
Reproduction, patterns of, 137–138
Reproductive cycle of the human female, 137
Reproductive isolation, 40
Respiratory system, 126, 130
 science practices, 151–153
Restriction endonucleases, 194
Reverse transcriptase, 195
Reverse transcriptase (RT), 200
Reverse transcription, 199
Ribosomal RNA (rRNA), 47
Ribosomes, 168, 191
Ribosomes of organelles, 105
Ribulose bisphosphate carboxylase (RuBP), 92
RISCs (RNA-induced silencing complexes), 141
RNA interference, 192
RNA polymerase, 166
RNA polymerase II, 192
RNA viruses, 200

RNA World Hypothesis, 54
Rough endoplasmic reticulum (RER) of organelles, 104, 121

## S

Saturated fats, 81
Sea anemones, 223
Seasonal reproduction, 70
Secondary structure, 83
Second messenger, 215
Secretin, 254
Secretory phase, 138
Selectively permeability of plasma membrane, 106–107
Selectively permeable plasma membrane, 111
Sexual reproduction, 70, 137
Sexual selection, 39
Shivering, 132
Short tandem repeats, 196
Shuttle step, 95
Sialic acid synthase B (SASB), 263
Sickle-cell anemia, 198
Signal peptide, 168
Signal transduction, 214–215
Signal-transduction pathway, 177
Single-celled organism, 47
Smooth endoplasmic reticulum (SER) of organelles, 105
Solid surfaces, 54
Solute potential, 112, 144
Speciation, 40
Species biodiversity, 266–267
Specified amino acid sequence, 167
Spermatogenesis, 137
Sperm cell, 139
Sphincters, 252
Spliced mRNA, 167
Splicosome, 166
Sporophyte, 135
Stabilizing selection, 38
Stereoisomers, 80
Steroids, 82
Stomata, 87–88, 275
Stratum, 55

Stroma, 87
Substance A, 127
Substance B, 127
Substitution mutations, 170
Substrate, 246
Sweating, 132
Symbiotic relationships, 255
Sympatric speciation, 41
Synaptic signaling, 213

## T

TACT mechanism (transpiration, adhesion, cohesion, tension), 275
TATA box, 166
Taxis, 278
Taxonomy, 51
Telomeres, 164
Telophase, 175–176
Template DNA, 167
Termination, 164, 169
Termites, 223
Terrestrial environments, 130–131
Tertiary structure, 83
Test-day checklist, 5
Tetraploid off spring, 41
Theory of endosymbiosis, 48
Thermolabile, 264
Thermophilic bacteria, 195
Thermoregulation, 128
Thylakoid membranes of chloroplasts, 88
Topic-level quizzes, 3
Topoisomerases, 163
Transforming factor, 158
Translation unit, 168
Transpiration, 136
 science practices, 275–277
Trichomes, 59–60
Triglyceride, 261
Triplet, 167
Trisomic chromosome, 198
Trisomy 21, 180
TRNAs, 167, 169
Trophic level, 239
*trp* operon, 127
Tryptophan, 127
Tumor suppressor genes, 194

# Index

Turgor pressure, 144
Type 1 diabetes, 129

## U
Uncatalyzed reaction, 246
Uniformitarianism, 30
Unsaturated fats, 81
Urey, Harold C., 54

## V
Vacuoles, 106
Valence electrons, 75
Vesicles, 105

Vestigial organs, 46
*Vibrio cholerae*, 215
Viral replication, 199–200
Vitamin A deficiency, 197

## W
Water molecule, 77–79
    hydrogen bonds, 77
    properties, 78–79
Water potential, 111–112, 136, 143, 275
Wax, 82
Webbing, 141
Whole plant transpiration, 276

Wilkins, Maurice, 159
Wisconsin Fast Plants®, 59, 271
*Wyeomyia smithii*, 37

## X
Xylem, 275

## Y
Yeast, 98

## Z
Zygote, 139, 230

# NOTES

## NOTES

# NOTES

# NOTES

# NOTES

## NOTES

# NOTES

# REA's Test Preps
## The Best in Test Preparation

- REA Test Preps are **far more** comprehensive than any other test preparation series
- Each book contains full-length practice tests based on the most recent exams
- **Every** type of question likely to be given on the exams is included
- Answers are accompanied by **full** and **detailed** explanations

REA publishes hundreds of test prep books. Some of our titles include:

**Advanced Placement Exams (APs)**
Art History
Biology
Calculus AB & BC
Chemistry
Economics
English Language & Composition
English Literature & Composition
European History
French Language
Government & Politics
Latin Vergil
Physics B & C
Psychology
Spanish Language
Statistics
United States History
World History

**College-Level Examination Program (CLEP)**
American Government
College Algebra
General Examinations
History of the United States I
History of the United States II
Introduction to Educational Psychology
Human Growth and Development
Introductory Psychology
Introductory Sociology
Principles of Management
Principles of Marketing
Spanish
Western Civilization I
Western Civilization II

**SAT Subject Tests**
Biology E/M
Chemistry
French
German
Literature
Mathematics Level 1, 2
Physics
Spanish
United States History

**Graduate Record Exams (GREs)**
Biology
Chemistry
General
Literature in English
Mathematics
Physics
Psychology

**ACT** - ACT Assessment
**ASVAB** - Armed Services Vocational Aptitude Battery
**CBEST** - California Basic Educational Skills Test
**CDL** - Commercial Driver License Exam
**COOP, HSPT & TACHS** - Catholic High School Admission Tests
**FE (EIT)** - AM Exam
**FTCE** - Florida Teacher Certification Examinations
**GED**
**GMAT** - Graduate Management Admission Test

**LSAT** - Law School Admission Test
**MAT** - Miller Analogies Test
**MCAT** - Medical College Admission Test
**MTEL** - Massachusetts Tests for Educator Licensure
**NJ HSPA** - New Jersey High School Proficiency Assessment
**NYSTCE** - New York State Teacher Certification Examinations
**PRAXIS PLT** - Principles of Learning & Teaching Tests
**PRAXIS PPST** - Pre-Professional Skills Tests
**PSAT/NMSQT**
**SAT**
**TExES** - Texas Examinations of Educator Standards
**THEA** - Texas Higher Education Assessment
**TOEFL** - Test of English as a Foreign Language
**USMLE Steps 1,2** - U.S. Medical Licensing Exams

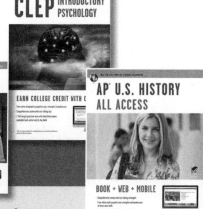

For our complete title list, visit www.rea.com

**Research & Education Association**

# REA's Study Guides

## Review Books, Refreshers, and Comprehensive References

### Problem Solvers®
Presenting an answer to the pressing need for easy-to-understand and up-to-date study guides detailing the wide world of mathematics and science.

### High School Tutors®
In-depth guides that cover the length and breadth of the science and math subjects taught in high schools nationwide.

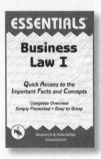

### Essentials®
An insightful series of more useful, more practical, and more informative references comprehensively covering more than 150 subjects.

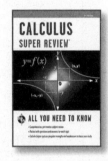

### Super Reviews®
Don't miss a thing! Review it all thoroughly with this series of complete subject references at an affordable price.

### Interactive Flashcard Books®
Flip through these essential, interactive study aids that go far beyond ordinary flashcards.

### Reference
Explore dozens of clearly written, practical guides covering a wide scope of subjects from business to engineering to languages and many more.

For our complete title list,
visit www.rea.com

**Research & Education Association**